DEEP OAKLAND

DEEP

OAKLAND

HOW GEOLOGY SHAPED A CITY

ANDREW ALDEN

Illustrations by Laura Cunningham

HEYDAY, BERKELEY, CALIFORNIA

Library of Congress Cataloging-in-Publication Data

Names: Alden, Andrew (Geologist), author.
Title: Deep Oakland : how geology formed a city / Andrew Alden.
Description: Berkeley, California : Heyday, [2023] | Includes
    bibliographical references and index.
Identifiers: LCCN 2022033094 (print) | LCCN 2022033095 (ebook)
| ISBN 9781597145961 (hardcover) | ISBN 9781597145978 (epub)
Subjects: LCSH: Geology—California—Oakland. | Oakland
    (Calif.)—Historical geography. | Oakland (Calif.)—History. |
Landscapes—California—Oakland.
Classification: LCC QE90.O2 A43 2023 (print) | LCC QE90.O2
(ebook) | DDC
    557.94/66—dc23/eng20221209
LC record available at https://lccn.loc.gov/2022033094
LC ebook record available at https://lccn.loc.gov/2022033095

Cover Art: Soderberg, Fred. "Oakland, California, 1900."
    Oakland: F. & H. Soderberg, c1899.
Cover Design: Archie Ferguson Design
Interior Design/Typesetting: Archie Ferguson Design

Published by Heyday
P.O. Box 9145, Berkeley, California 94709
(510) 549-3564
heydaybooks.com

Printed in East Peoria, Illinois by Versa

10 9 8 7 6 5 4 3 2

FSC
www.fsc.org
MIX
Paper from
responsible sources
FSC® C005010

To John Robert Alden

CONTENTS

PREFACE

People are naturally drawn to natural things. Given a chance, they're ready to listen to birds, to reckon trees, to attend to all kinds of living creatures. Many are called by the clouds and winds and stars. Fewer are drawn downward: listeners to landscape, reckoners of rocks. Nevertheless, we're all shaped by the geological as well as biological forces of the Earth.

Every city sits where it does for geological reasons, which may include the availability of underground and surface water, ready access to natural sources of wealth, terrain suitable for rail lines and highways, navigable rivers and ports for shipping, amenable climate and exceptional scenery.

Regarding a city in terms of its geology—seeing it from the ground down—is not the usual way. Until well into the twentieth century, geologists weren't consulted before cities were founded. But sooner or later, growing cities hit limits imposed by their geology, and how they respond affects whether they thrive or founder. The longer I explore

Oakland, the more I see how its unusually rich geological setting steered its history and constrains its prospects.

Beyond Oakland's value as a case study, if you're here, you're in the midst of an underappreciated teaching tool, a textbook for a wide range of geological concepts. It has features its managers might better heed and others its residents might treasure.

When I say the word *geology*, most people think I mean rocks, but it's about much more: the work of streams, the formation and movements of sediment, the rise and fall of mountains, the changes of living things as age succeeds age. By geology I really mean how the Earth works—how *planets* work. Geology is the study of worlds made of rocks. The Earth is active, geologically alive, and every place on it takes part. Geology opens one's eyes to a secret world that lies beneath and all around us. Knowing some geology is like having a key to the house that opens up a vast extra room. Here I hope to introduce the ways of Earth to you.

Why does Oakland geology matter? Every town and city has its own geological setting. Oakland is a special city that owes its origin and character to a remarkable setting, making it not just a beautiful place but a good one to learn the basics of geology. The Oakland Hills tempt me every day to visit them, and when I do they dazzle me with views as well as rocks. By my estimate, Oakland has more kinds of rock—red, white, blue, and green—than any other city in America. That's my idea of a tourist attraction. Once acquainted with the subject, anyone can turn to their own hometown and see it with new eyes. Turning on one's geologist eyes is something like snorkeling: the surface world with all its sound and motion drops away, and an entirely different world swims into view.

...

Oakland has always been praised as a beautiful place, its range of high hills descending in a wave to a wide coastal plain. But since the day the original inhabitants, the Ohlones, were joined by outsiders in 1770, this place has differed with each set of eyes that beheld it. The Americans who founded the city exploited the natural resources of its core territory from the start: timber, firewood, water, soil. As Oakland grew, they found more natural assets in the surroundings to turn into money: clay, stone, gravel, seafood, sulfur. Each of these resources has a connection with our distinctive geology. Each in its turn was consumed, gratefully but imprudently, until it was exhausted or made uneconomical by sources in more distant places. The local resources that survive—the landscape, the views, the light, the air—are inexhaustible and precious, and these too have their place in the geologist's view of Oakland.

It's a truism among people who work on sustainability that our actions should consider the effects on the seventh generation to come. Oakland was founded in 1852, and for me that's close enough to seven generations ago. Therefore it's time to consider the consequences of the city's first decisions as we look seven generations ahead.

Oakland has gone through four epochs of human history, which I define here in strictly local rather than regional or national terms. The Ohlone period, named for the tribal peoples who were the first to belong to this land, began late during the last ice age and ended in approximately 1800. The Spanish period began locally with the forced removal of the Ohlones, shortly after the first expeditions from New Spain in the late 1700s. The Mexican period began with independence from Spain in 1821. For convenience, I refer to US citizens and residents as "Americans." The American period began in 1850 and continues to the date of this writing. Oakland is an American city, but it contains significant populations representing all three previous periods that make up part of our special sauce.

The history of California is marked with violent acts. They're important to recognize. Briefly put, the Spanish crown basically enslaved the local tribes and separated them from their land under the auspices of the Catholic church. The government of Mexico then dispersed their land to private hands, spawning a short-lived rural aristocracy of Mexican ranchers whose free-range livestock had it better than the remnant tribespeople who worked for them unpaid. The United States seized it all as a war prize, turned the best parts into gold and neglected whatever didn't measure up to that standard. The original peoples of California were subjected to actions, both legal and extralegal, that constituted genocide. In Oakland, similar actors at the local level carried out similar acts, especially within the flexible bounds of nineteenth-century governance and justice.

The story of California rests on white entitlement, a virus from colonialist Europe that, with each expansionist wave westward, mixed genes with capitalist entitlement. Every age has its conquerors and its mercenaries. Just as the three squatters who founded Oakland made their initial underhanded move under a pretext that the American takeover of 1848 erased all previous land titles, so geologists— my people—were the paid servants of expansionism in exploiting public lands, erasing their native overseers' claims, for private profit and national glory. Geologists armed with academic credentials, traditionally persons male and white, participated in a centuries-long wave of exploration that began in the 1830s with the first state geological surveys and swept west with the flag and telegraph and railroad, then overseas with the American empire. They are my scientific ancestors. It was a glorious age of scientific progress, borne on the horses of crushing imperial conquest.

Oakland was founded and built by men of that imperium, self-assured men, ruthless by upbringing, men who sought an advanced, idealized life for themselves worthy of the particular kinds of natural wealth and beauty they perceived. Among geologists in the generation now gone, I used to see a sort of lingering cowboy privilege, a pride that I think was inherited from those days in the vanguard, roaming and plundering the western frontier. We were on the winners' side of imperialism.

Yet our science is also sensually rich and intellectually sublime. I think that by the last chapter this will be plain. The geologist looks for principles in the swarm of particularities that make up landscapes and regions. The point of finding the scientific laws of rocks and landscapes is to apply them, using those laws to gain insight into every place on Earth, each in its own particularity. The aim of geology is to help us awaken tomorrow and know our home ground afresh. The fact that geologists can have jobs of mundane routine and careers of applied expertise does not erase their dream.

Readers who aren't geologists may not notice that I use geological terms that look like everyday words. *Clay*, *silt*, *sand* and *gravel* are size classes of sediment particles, but in everyday terms clay is creamy, silt looks smooth but chews gritty (yes, we taste rocks), sand is grainy (from barely visible to millet size) and gravel is larger than pea size. *Mud* is another of these terms; geologists use it to mean a mixture of clay and silt.

Readers who are geologists may notice that I use some geological terms in this book without the rigor they might prefer. For instance, I refer to the segments of the geologic time scale as "periods," although working geologists are particular about using the correct time rank in the nested series that consists of ages, epochs, periods, eras and eons. To them I say: believe me, as someone who's made a living enforcing our terminology, it was hard to relax.

I speak in this book as a member of the geological community, where I have spent my working life since college in a supporting role. When I speak of the attitudes and attributes of "my tribe," I reflect only my own opinions based on long experience and do not speak for anyone else. Neither do I wish to pretend that my community is truly tribal.

Before I start, a few orienting notes. The East Bay is a belt of territory, coastal flats in front of a range of high hills, that is generally aligned north–south with a distinct leftward lean of about 32 degrees. People here ignore that lean when they talk about directions—they go north to Richmond and south to Fremont. But Oakland lies on a wider lump of East Bay land that sticks out into San Francisco Bay, such that downtown is decidedly west of the southern end of the city. Southern Oakland has been called East Oakland since the early days. The traditional division between East Oakland and the rest of Oakland is at Lake Merritt, an arm of the Bay that intrudes deep into the coastal plain.

Much of Oakland is covered by a system of numbered streets, and new residents quickly learn that the Avenues, 1st through 109th, are in East Oakland running perpendicular to the coast, while the Streets, 2nd through 66th, lie west of Lake Merritt and run parallel to the shore. (East 8th through East 34th Streets extend the scheme partway into East Oakland.)

Oakland shares California's Mediterranean climate: a seasonal cycle, unique in the United States, of rainless summers and cool, wet winters. Our landscape turns gold in summer as the grasses die back, then green with the late fall rains. Most native trees, but not all, are evergreen species that never shed all their leaves. Snow in Oakland is rare enough to make the evening news; so is heat above body temperature.

• • •

My experience of Oakland, my city for over thirty years, is one of growing intimacy with its unusual geological complexity. I have walked every block of its streets, inspected all of its stairways and trails, and done a fair amount of off-trail bushwhacking. Oakland is a mosaic of long-gone lands—lost worlds—some of which have traveled great distances, over great stretches of time, to spend time with us. Its parts have been gathered and sculpted, by the aimless forces that operate the Earth, and stitched into a landscape rich in personality. Geologic Oakland is the core of my home and the navel of the world.

Every geologist has what might be called a natal landscape: the kind of country they lived in as a child, the landscape they imprinted on. The Bay Area's hills and peaks are the first mountains I ever saw. As with my father before me, my natal landscape is the Bay Area; but also like him I have deep family roots in the Northeast and spent important parts of my life there. Landscape was my childhood fascination, and it still is. I was raised on both the East and West Coasts, but not until my geology degree sank in did I see and understand their profound differences. The East Coast, in plate-tectonic terms, is a passive continental margin: for over one hundred million years, since the Atlantic Ocean opened up, it has stayed quiet long enough for erosion to do its work, keeping the coastal plains flat and the mountains fairly low. The West Coast is an active margin, where tectonic forces have cut and rearranged the Earth's crust in dramatic fashion and where earthquakes remind us it's still on the move. Thus this book begins with the Hayward Fault and ends with the Oakland Hills, which owe their existence to it. Every California city has a view of mountains, but our hills are special, as the Ohlones still living among us have always known. Come see.

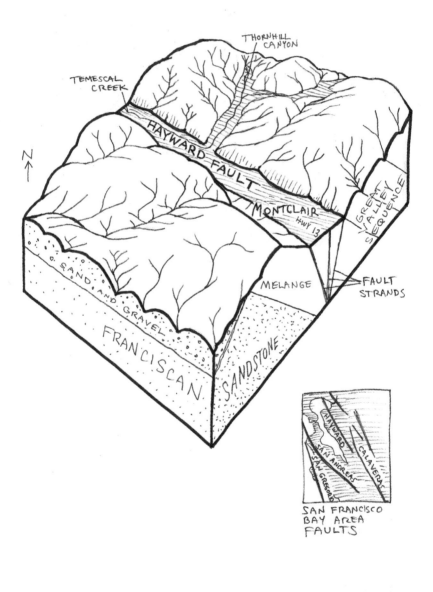

THORNHILL CANYON

TEMESCAL CREEK

HAYWARD FAULT

MONTCLAIR
HWY 13

GREAT VALLEY SEQUENCE

N

SAND AND GRAVEL

FRANCISCAN

MELANGE

SANDSTONE

FAULT STRANDS

HAYWARD
CALAVERAS
SAN ANDREAS
SAN GREGORIO

SAN FRANCISCO
BAY AREA
FAULTS

1

THE HAYWARD FAULT

The key to Oakland's unusual landscape is the Hayward Fault. It threatens us with destruction by earthquake, yet our city would not be Oakland without it. Earthquakes are how the fault brought together our special place, and earthquakes are what maintain it. Oakland is earthquake *country*. The fault has made us beautiful, but it also keeps us on our toes.

The Hayward Fault runs north–south for fifty miles, from the shore of San Pablo Bay to a point east of San Jose. It's not just a line on the map but an underground surface shaped like bedsheets waving on a clothesline. It extends downward to a depth where the rocks are too hot and soft to crack, about ten miles. It's the foundation of the East Bay's landscape, a major seismic feature, a scientific attraction in our midst. It's also an intermittent catastrophe, bad for business, a geological embarrassment. Wherever it goes, the people all complain. Just mentioning the Hayward Fault can feel a bit indecent.

A fault is what geologists call any crack that has allowed the rocks to move along its surface, like hands rubbing together. Geologists see faults everywhere, and nearly all of them are inactive, even the ones big enough to mark

1

in black lines on geologic maps. They last moved long ago. The Hayward Fault is an active fault, busy storing energy and releasing it in earthquakes. Active faults are entities defined in California law, with names and histories and public reputations.

The fault is responsible for the Oakland Hills, though the connection can be hard to see through their familiar changing appearance. In spring the hills evoke green bedclothes on a sleeper. In summer they wear motley gold and olive, in winter a robe of deep green-gray. At dawn they present a uniform dark mass; at midday their laid-back slopes invite a siesta. The afternoon light brings out their sculpture; sunsets paint them bronze. Viewers with high imaginations may perceive the Oakland Hills as a massive wave.

Such a moment offers a glimpse of the geologist's view, in which these hills are not just beautiful features but dynamic ones. They really *are* a wave of rock, rising in motion measured in millimeters per year, centimeters per decade, meters per millennium. This is tectonic motion—movement that involves the whole landscape and the full thickness of the Earth's crust. The Oakland Hills are thought to have started rising roughly a million years ago. To a geologist, they're quite young.

The geologist's view is unmoored from ordinary time. When I refer to faults moving and hills rising, I picture smooth motions taking place over thousands of years. But in ordinary time, that motion generally takes place in discrete jumps of a meter or two that trigger large earthquakes. These happen at irregular intervals, whenever enough stress accumulates to overcome the friction holding the fault still.

The Hayward Fault is one of a family of parallel faults that slice and dissect the whole Coast Range. By rearranging the Earth's crust over millions of years, one earthquake at a time, these faults have brought bodies of rocks that were once far apart into odd juxtaposition. And they link us to a global web of seismic activity called plate tectonics, the system that stirs the Earth's outer layer in continent-sized pieces.

Oakland includes more of the fault's length than any other East Bay city. I've made it my business to visit every accessible bit in Oakland. We don't have to love the Hayward Fault, but I believe we should all recognize it and understand how it shapes Oakland's landscape and the city's future.

At the time I write this, the Hayward Fault hasn't shattered the ground since the fall of 1868, but it will again. On that day, everyone will see it. But today the fault is a latent presence with subtle signs, visible only with guidance and practice. To those who seek it out, note that there's something weird and suspenseful about seeing the fault today. That feeling doesn't go away with scientific training.

At the north end of Oakland is the Claremont Resort, a great white palace built in 1915 at the foot of the Oakland Hills on Claremont Avenue. It sits near Claremont Creek at the mouth of a dramatic gap in the hills named Claremont Canyon. The developers who branded everything "Claremont" were attracted by the majestic terrain where the canyon cleaves the hills. The Hayward Fault uplifts those hills and defines that terrain, and it intersects the Earth's surface mere yards behind the resort. It's easy to see the fault's vertical influence here.

Stonewall Road, across the canyon's mouth from the resort, heads north directly upon the fault, and the topography is stark. Downhill the land slopes mildly leftward toward the Bay, but uphill on the right the slope is dizzying, truly a stone wall. The street takes a hairpin turn to the right and rises steeply up this wall toward a dead end. The view grows wider with each step.

When I look west from here over the North Oakland plain, sometimes I can't stop a surge of dread as I picture the next major earthquake and its fearsome aftermath: hundreds of structures shaken to bits, thousands more damaged, uncountable lives thrown off track. I foresee Stonewall Road completely blocked by wreckage, rockslides and fallen tree limbs—I see scores of fires break out below—they spread and merge in walls of flame—the firestorm closes with a roar, but there's no escape, no escape!

I know why people prefer not to think about the fault. Every time I visit it, no matter where, I push down the dread. But once I do that, there is room for wonder and interesting things to see.

As I look southward across the mouth of Claremont Canyon, the opposite side looms near, surprisingly close and steep. This isn't how stream valleys are supposed to work. Stream valleys are supposed to grow wider and gentler as they descend from their headwaters, not narrower and steeper. The Hayward Fault causes this distortion as the forces acting on it raise the Oakland Hills.

Although the fault's two faces move past each other mostly sideways, a small amount of compression across the fault pushes up the east side by a millimeter or so per year; that's why there are hills here. Geologists call this blend of forces transpression. As the threshold of Claremont Canyon rises, Claremont Creek cuts down hard into it, and the creek has carved the canyon's mouth into a deep notch with steep walls. Once across the fault, the creek spreads wide upon the open plain in times of exceptional flood, delivering mud and gravel in all directions to build what geologists call an alluvial fan. Thus the fault, over the million years or so that the hills have existed, has turned Claremont Canyon into a shape called a wineglass valley—the wide headwaters form the bowl, the narrow notch the stem, the alluvial fan the base. Claremont Canyon is the clearest example of a wineglass valley on the Hayward Fault, but most of the canyons in the Oakland Hills share its features to some extent.

Features like alluvial fans and wineglass valleys, created by vertical motions of the Earth's crust, are common all over the American West. In fact, from the Sierra Nevada all the way across the Rocky Mountains, most of the active faults move vertically, and the mountains have grown much higher there. In the Bay Area, vertical motions are more modest. The real action around here is horizontal, what geologists call transcurrent motion. On the Hayward Fault, the transcurrent motion is much greater than the uplift, about nine millimeters per year on a long-term average. Its west side moves north with each large earthquake, making it what geologists call a right-lateral fault.

• • •

A good place to see features created by the Hayward Fault's right-lateral motion is at Lake Temescal Regional Park, a former reservoir, made by damming Temescal Creek, that's been popular for swimming and fishing since 1934. The lake sits in an unusual valley, a narrow canyon that cuts not west down the slope of the hills, like Claremont Canyon, but north across the hills, following the fault. Valleys like this are common along active faults in the Coast Range, because thousands of repeated earthquakes grind and shatter the rocks near the fault. This pulverized material readily turns to nutrient-rich soil; it's also easily eroded, and therefore faults are places where streams can dig deep. As one observer put it, "An earthquake fault is often marked in California by successions of dairies and of reservoirs."

The west side of the lake is a high ridge of bedrock. It extends southward out of sight, but just north of the dam it abruptly ends. Temescal Creek exits the dam north through a spillway, then takes a hard left turn around the end of this ridge and flows west to the Bay. The ridge is being carried north, and over the centuries it has forced the stream to flow more than a mile northward from its source in Thornhill Canyon. Geologists call landforms like this shutter ridges, after the shutter of an old-fashioned camera. Others in Oakland, and many more on the rest of the fault, have put kinks in the East Bay streams over tens of thousands of years.

This canyon tempted entrepreneurs in Oakland's early days to build a dam here to supply the town with running water. First to succeed, in 1869, was Anthony Chabot with his Contra Costa Water Company. Chabot's dam is a wide pile of packed soil with a thick core of clay, and even though the Hayward Fault runs right through it, engineers consider it safe in any foreseeable earthquake.

South of the lake, by the ranger office, an interpretive sign directs attention to the very fault itself: a set of short cracks in the asphalt path. The cracks are arranged like the steps on an escalator, lined up along the track of the fault with each crack pointing aslant to it. This arrangement is called by a French term, *en echelon*, that geologists pronounce "en-*esh*-a-lon." Crack sets like these are telltales of the fault everywhere in the East Bay, but this is the only place in Oakland where they're thought fit to show the public.

En echelon cracks are products of a process called aseismic creep. Some faults move between earthquakes softly and quietly, a few millimeters per year. The Hayward Fault creeps, and is famous for it. Faults that creep are rare enough that geologists travel here to witness one so well exposed.

In repeated visits, I've watched the cracks at Lake Temescal lengthen and open. Nearby is an underground chamber where a wire thirty meters long stretches across the fault between two concrete piers. Records from this creepmeter show that the fault silently moves about three millimeters a year.

Between big earthquakes, creep works its influence on Oakland neighborhoods. The signs include poorly fitting doorways, cracked pavements and the occasional leaking water main, or spots where these things have been repeatedly repaired. A common sign is bent curbs, where the sidewalk's edge, over the space of a few yards, veers a few inches to the right and then curves back.

On the south side of 39th Avenue, high up by its top at Victor Avenue, the concrete curb takes a little skip where it crosses the fault—not a gentle bend but a clean sharp shift. Someone took a drill with a rotary blade and made a little sawcut across the offset curb segments. Over the years the two parts of the cut have kept on moving to each other's right, millimeter by millimeter. I come back here again, every now and then, and take a picture.

Researchers come here on a schedule to measure the positions of little pegs laid in a long row across the fault, using a surveyor's theodolite on a precision tripod. They've repeated this exercise for decades at a couple dozen localities like this. On that painstaking bit of basic science rests much of what we understand about creep on the Hayward Fault.

Suitable places for these measurements are hard to find. Dirt muffles the signals of creep, and earthquake ruptures are soon erased by rain and vegetation and gophers. Humans are even better at hiding the signs, especially where a fault trespasses on their property. For instance, on Stonewall Road, I watched the pavement crack for years until the city repaved the street. Geologizing can be challenging in an urban setting. This spot on 39th Avenue is undisturbed because no resident owns it.

The same fault can both creep and lurch because of how faults work. As the two faces of the fault surface are pulled past each other and friction restrains them, the Earth's crust deforms, like rubber blocks. The rocks store elastic energy, like the rubber in a rubber band, and release it when the fault gives way in an earthquake. The energy is greatest down deep in the fault, where the weight of overlying rock holds the two sides together hardest. The biggest fault ruptures, which cause the largest earthquakes, happen down there. Creep, by contrast, happens only in the top kilometer or two, where the overlying load is lighter and little friction energy builds up. Creep doesn't affect the deep crust where large earthquakes arise. Although creep may respond to an earthquake by changing its pace, it doesn't greatly influence future earthquakes. And the landscape can't tell the difference: it evolves the same way whether a fault creeps or not.

Indigenous tribes in California explain earthquakes as movements of the seven giants who hold up the world. As it happens, seven giant tectonic plates account for more than 90 percent of the world's surface area, so Indigenous geology fits well with modern understanding. So do the traditional Indigenous earthquake-resistant lifestyles. The Ohlones lived in reed huts and built nothing larger than a sweathouse or a small meeting hall. Large earthquakes always disrupt the land, stressing the societies that depend on it.

But surely few Indigenous Californians ever died in earthquakes until forty of them perished when the earthquake of 8 December 1812 demolished the new stone chapel at the San Juan Capistrano mission during morning Mass.

The Bay Area was seismically busy during the nineteenth century, more so than today. A damaging earthquake in 1808, recorded at the San Francisco presidio, was the first of a dozen major events during the century before the great earthquake of 1906. The biggest one of that century was in 1868, and the Hayward Fault was the culprit.

In Montclair Park, near Oakland's midpoint, the fault is invisible on the grassy ground, but scientists uncovered its latent trace there in 1994. By the baseball diamond in the park, just past third base, Jim Lienkaemper of the US Geological Survey led a team that dug a trench fifty feet long across the fault. A trenching study is exploratory surgery in the ground, carried out with forensic precision.

Lienkaemper is a quiet, meticulous researcher with sharp eyes and soft hands. Working out of the Survey's regional office across the Bay, he's been in more trenches than anyone else in Northern California, perhaps the world. I joined him one day in a trench across the Green Valley Fault, north of the Bay in a meadow west of Fairfield. A backhoe had dug out a slice of the clay to a depth of about six feet, just wide enough to squat in. Steel braces propped up the trench's vertical walls. The research team had carefully shaved the walls clean, laid on them a grid of strings, photographed it all and filled their notebooks with sketches and observations. Lienkaemper moistened the walls with a spray bottle to bring out the colors and pointed out intricate features my unpracticed eye could barely perceive, their boundaries marked with tacks and string. Scratches outlined a crumb of charcoal to be sampled for radiocarbon dating. Nearby a plywood panel held a mosaic of photos, spiderwebbed with annotations.

Lienkaemper's trench in Montclair Park showed that the ground broke during the October 1868 earthquake. The "great San Francisco earthquake," as it was known until the April 1906 San Francisco earthquake, occurred when the southern half of the Hayward Fault ruptured the surface from Fremont to here in Montclair and possibly beyond. No one alive has felt a quake this strong in the East Bay.

• • •

The year 1868 came in an edgy decade. In the previous nine years, four major earthquakes had shaken the East Bay, the latest in 1865. The newspapers had noted smaller, local shocks on 24 March and 24 August. Within the latest twelve months, catastrophic earthquakes followed by giant tsunamis had struck the Virgin Islands, Hawaii and the coast of Peru—and in all three places they're still the largest ones on record. The newspapers that week were still mentioning them.

The Hayward Fault shrugged on 21 October, a foggy Wednesday morning, a few minutes before eight o'clock local time. The sides of the fault shifted sideways as much as two meters. Shaking was felt for hundreds of miles, from Chico to Monterey and east into Nevada. The air shuddered with sounds like thunder. Dozens of aftershocks occurred that day, and hundreds more followed within a week.

Local newspapers reported that the ground "rolled with huge waves like the sea" and looked like "the backs of a drove of buffalo at full speed." A worker in Alvarado said, "The mountains seemed to skip like." In more distant Sacramento, waves on the river made boats "surge to and fro and champ at their hawsers."

In Oakland, "trees whipt about like straws." Streams flung water over their banks; milk pails were splashed empty. Cracks broke the ground in a ragged line stretching about twenty miles from present-day Fremont north past San Leandro and, by some accounts, into southern Berkeley. Some of these fissures threw water and sand high in the air.

San Leandro Creek, usually a trickle late in the dry season, turned within minutes into a foot-deep torrent, reported the *Alameda Democrat*, "rushing down the bed of the creek as turbid as though a violent rain storm had been raging in the mountains east of us." In Hayward, springs and wells shut off their flow or surged with excess water. In Fremont, Tyson's Lagoon drained and stayed dry for three years.

The shaking damaged much of San Francisco, including the city hall, and collapsing buildings killed about thirty people. East Bay victims included two people in Hayward and Joseph Josselyn, "a young man of much promise and ability," who died when the stone doorway of the new courthouse in San Leandro crumpled upon him. His fellow Masons buried him in Oakland's elegant new hillside cemetery with a procession of more than eighty coaches. Then as now, it was buildings, not earthquakes, that killed people.

The brick buildings of downtown Oakland suffered widespread damage, although the paper reported that the brand-new Wilcox building on Broadway at Ninth Street, the tallest structure in the city at three stories, "stood admirably. It is well braced with iron." It still stands admirably today.

But once the aftershocks faded away, so did any efforts to learn from the event or build back better. The San Francisco Chamber of Commerce appointed a committee to gather data and publish a report, but nothing came of it. The members were mostly amateurs; the secretary complained that their meetings had a "desultory conversational character." And when it happened that the committee's chair died five months later, the group disbanded.

A rumor persisted, basically an urban legend, that the committee finished a report that was suppressed because it looked bad for business. Like most urban legends, the rumor had a varnish of plausibility. Stakeholders did actively minimize the catastrophe, and the stakes were high. San Francisco, leading city of the West, had a reputation to protect. And in Oakland, the planned terminus of the nearly complete transcontinental railroad, any hesitation would threaten its future as a great commercial city.

With straight faces, the newspapers asserted that California earthquakes were actually mild, not like the notorious ones overseas, and the damages were trivial. The very next day, the *San Francisco Chronicle* proclaimed, italics and all, that "after a careful analysis of reports from every quarter we find that there *is not a single case where any well-constructed building, standing upon solid ground,* has been damaged." Apparently the threat could be simply defined out of existence.

More realistically, the Hayward Fault earthquake was much smaller than those in Peru, Hawaii and the Virgin Islands. It had a different tectonic mechanism, and thus it produced no tsunami—the waters of the Bay had stayed notably still. The day and hour meant a relatively small death toll, with schools and churches empty and many workers not yet inside the most vulnerable brick buildings. And absent the wholesale fires that made the 1906 earthquake so notorious, the booming economy made recovery swift.

After 1906, the older quake was so thoroughly forgotten that in 1930, the city of Hayward dedicated a handsome new city hall built on the 1868 rupture right where it had ripped apart the adobe home of Don Guillermo Castro. Creep on the fault soon pulled the new building askew, and it had to be abandoned. At this writing the Art Deco landmark still stands, vacant and useless. The first time I visited it, I peered in the windows at the bent banisters and cracked walls inside. Then I walked over to the road, and there they were, en echelon cracks. As I watched, a little tongue of water came out of one, leakage from a freshly broken main. I found a telephone booth, fed in some coins, and informed the public works department.

In Oakland in 1868, Montclair was an outlying place of no consequence. Local farmers must have seen the fissure recorded in Lienkaemper's trench; a little north, the country estate of Colonel Jack Hays suffered severe damage. Amid the rush of news during the following week—among other things, the first presidential election after the Civil War was days away—few took note of this remote district. The event was easy to minimize. The cracks in the ground soon healed, and fifty years later subdividers and developers moved in as if the fault didn't exist. The city made no fuss. And so today the fault lurks directly beneath the bustle of Montclair's commerce. Where it runs through a supermarket, it ripples the floor.

State law allows buildings that straddle an active fault to stay as long as they're kept repaired. But buildings damaged beyond repair cannot be replaced, and the lots they occupy will become permanently empty—properties condemned by nature. Within a century or two, as earthquakes repeat and empty lots accumulate, the Hayward Fault's surface trace will likely become a long greenbelt, a change from eyesore to amenity.

• • •

Whenever I visit the fault in person, it feels fine. Even in this part of the world, active faults are usually as quiet as a parked car, and seismic shaking is a sporadic thing. At home I feel the jiggles of minor events and go back to my business. A decent shaker, one that throws tchotchkes off shelves, might visit the Bay Area a few times per decade, about as often as a decent thunderstorm.

Still, feeling a fault move gets one's attention. As Mark Twain reported after the major San Francisco earthquake of 8 October 1865, "Usually I do not think rapidly—but I did upon this occasion. I thought rapidly, vividly, and distinctly." There's a *heave* to even a little earthquake that's not like doors slamming or trucks passing. Stronger ones are regional events that trigger car alarms everywhere in earshot. The biggest, history tells us, can shake one's faith in civilization.

But the Earth is no longer an unfathomable realm. In my lifetime, within my own memory, the theory of plate tectonics has given us our first comprehensive understanding of the planet. This scientific framework treats the Earth as an outer shell made of large moving segments, as if the polygons of a soccer ball could be manipulated like a Rubik's Cube. Moving at the speed fingernails grow, these tectonic plates expand, collide and are consumed as they circulate over millions of years.

Plate tectonics explains the Hayward Fault as one of a sheaf of faults where the Pacific and North America plates scrape past each other. Earthquake geology, for centuries a field of naive speculation, is now well constructed and on solid ground. Plate movements are measured with millimeter-per-year accuracy, earthquakes are located with growing speed and precision, and researchers are having a lively and fruitful conversation.

We're learning more about the Hayward Fault's deep

shape. The fault is continually busy with earthquakes too small to feel. The size of an earthquake, that is, the amount of energy it releases, depends on the size of its rupture. Seismometers can detect waves from a rupture as small as a pickleball court and fix its three-dimensional location—its hypocenter—with similar accuracy. Between the ground and the hot depths where the rocks are too soft to fracture, thousands of hypocenters define a wavy surface like that of a theater curtain being drawn shut.

Near the south end of the fault, east of San Jose, its underground surface billows eastward, toward the nearby Calaveras Fault. There the two faults join and the Hayward Fault loses its identity. At its north end, the fault surface bends eastward beneath San Pablo Bay and connects with the Rodgers Creek Fault, which continues north to Santa Rosa. Given enough energy, a growing earthquake rupture can punch past either end of the Hayward Fault and rip into these adjoining faults. It may be that on the large scale, the three faults are effectively one. The ways we define and name faults may hinder our vision.

We're learning more about the fault's deep history from studying the rocks around it. Over the last twelve million years, the rocks of the greater East Bay have been offset a total of some 175 kilometers, of which about half occurred on the Hayward Fault as it is mapped today. Activity migrated among several other faults, some of which are inactive today, and the rates of motion varied on each one at different times.

We're learning more about the fault's recent history of large earthquakes by digging more trenches. Jim Lienkaemper and his colleagues have determined that the last dozen large quakes on the Hayward Fault broke the ground about every 160 years, on average, plus or minus many decades.

Although the time elapsed since 1868 is nearing that average, the cold science of statistics says that the odds of a big quake on any given day are not getting appreciably higher. The odds of one happening within the next few decades are somewhere between a roll of a die and the flip of a coin. Still, each day brings the next big one closer. Energy accrues, and the next big one may—*may*—be growing bigger as it waits to be born.

When the next one comes, there will be damage and permanent loss. Well-constructed buildings on solid ground are important, and steady improvements in California building codes have saved many lives. Today we also understand the concept of societal resilience, the notion that well-prepared, well-rehearsed cities and regions can take all kinds of disasters in stride. And afterward, Oakland could grasp the opportunity not to return to the old normal but to establish a new one: not to rebuild what used to be, but to build what should be.

Meanwhile, Oaklanders live their lives today in homes and buildings that sit on the fault. One afternoon I took some visitors to a suburban tract at the south end of Oakland, on a level patch of land backed by steep, oak-studded hillsides. At the base of the hills, where the Hayward Fault runs, Revere Avenue is riddled with en echelon cracks and dislocated curbs before it hooks left and climbs to a dead end in the hills, a mirror image of Stonewall Road.

I was showing these cracks and curbs to a pair of French journalists, a writer and a photographer. I try to be discreet when I make these visits, I told the two women, because residents may not appreciate the enthusiasm of random strangers. They found the fault interesting, but they were also here to explore the dread. Whereas most of the United States has experienced large earthquakes, France is seismically fairly quiet outside its bordering mountains, the Alps and the Pyrenees.

When a resident stepped out of a house that straddled the fault, the writer approached him. He admitted with a shrug that the fault could well destroy his home, but it's been fine so far. She returned to me, taken aback, and said, "We don't have attitudes like this in France." As a Californian, I had to reply that we've been this way since before the Gold Rush. The land taught the Ohlones long ago that dwellings and possessions are as dust.

One day the fault will reawaken beneath the places I've mentioned, maybe all of them at once. Californians have always made a show of taking earthquakes in stride. What else can we do? Because cracking ground will kill very few, if any, people outright, one can live upon the Hayward Fault with the right mix of denial, preparation and equanimity. I would never do that—my place three kilometers away is close enough—but people do. How can I, an Oaklander myself, condemn them for living in earthquake country? This town is not geologically safe, but then, few California cities are. The important thing is to make Oakland a resilient city, living *with* our fault and ready to take its earthquakes in stride.

Meanwhile, there's far more to the Hayward Fault than the immediate threat it poses. Patiently over millions of years, it has arranged Oakland's landscape, affecting all of its parts—and the lives of those in them—in ways it takes a geologist to appreciate.

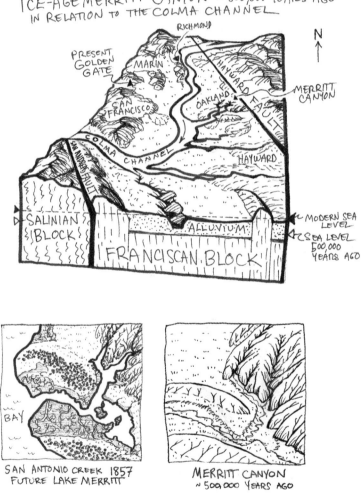

ICE-AGE MERRITT CANYON ~ 500,000 YEARS AGO IN RELATION TO THE COLMA CHANNEL

RICHMOND

N

PRESENT GOLDEN GATE

MARIN

HAYWARD FAULT

MERRITT CANYON

SAN FRANCISCO

OAKLAND

COLMA CHANNEL

SAN ANDREAS FAULT

HAYWARD

MODERN SEA LEVEL

ICE SEA LEVEL 500,000 YEARS AGO

SALINIAN BLOCK

ALLUVIUM

FRANCISCAN BLOCK

SAN ANTONIO CREEK 1857
FUTURE LAKE MERRITT

BAY

MERRITT CANYON
~ 500,000 YEARS AGO

LAKE MERRITT

Whereas the Hayward Fault perturbs us with energy from the Earth's interior, Lake Merritt connects us to the world ocean, the world atmosphere and the cosmic cycles of the solar system. Unlike the fault—invisible, inexorable, indifferent—the lake is open to all of Oakland, an intimate civic partner that registers and responds to every human touch. The landscape around Lake Merritt, near Oakland's historical core, is rich with geologic features dating from a million years before the city was founded. Although a century-plus of human action has thoroughly changed the lake itself, the landforms around it are still plain to see.

Geologists see landforms the same way they see rocks, fossils and faults: not just as features but as puzzles, clues to stories of how the Earth works. Lake Merritt's story is a combination of two stories, each a million years long: the repeated cycles of the sea during the ice ages and the steady rearrangement of the land by the Hayward Fault. This chapter tells the sea's part, and later chapters will tell the rest.

Our lake is a world-class oddity, an arm of the Bay in the midst of a city. It rises and falls with the daily tides. An inside-out island, a marine habitat surrounded by land, it is truly a mediterranean sea. In Oakland's early days it was a barrier, always in the way, but we've come to embrace it as the centerpiece of the room, framed like a sculpture, the city's focal point.

But Lake Merritt is not really a lake. The early English-speaking visitors called it a slough—a tidal backwater with slow-moving currents and muddy banks—and named it for the Rancho San Antonio, the Spanish private land grant that once extended from Berkeley to San Leandro. Dredging has turned the San Antonio Slough to open water, but it remains an estuary, a brackish zone between fresh and salt water. Compared to lakes, estuaries are dynamic places, prone to waves and tidewash, cloudy with microscopic life and mineral nutrients, where a riot of species thrive in mingling waters.

The City of Oakland owns Lake Merritt. Its entire shore is a city park with attractions for people watchers, bird watchers, garden lovers, exercisers and more. Circling the lake on foot is a popular thing to do, a circumambulation of about three-and-a-half miles, or six kilometers to a scientist. The geologist, ever attentive to landscape, notices that different types of terrain around the lake give it a visual rhythm, part of what keeps it so interesting.

On the map, Lake Merritt has the outline of a lower-case letter Y, defined by a long fat stroke slanting northeast–southwest and a short narrow stroke extending northwest to the mouth of Glen Echo Creek. But it's important in geology to distinguish the human imprint from nature's handiwork. The earliest good map of this area, published in 1857, shows the slough continuing southwest to the Bay in an open channel as wide as the rest of the lake. The long stroke of the Y was once twice as long as it is today. The slough's natural shape is that of a person's forearm, thumb raised as if to shake hands, truly an arm of the sea in a gesture of welcome. The wrist of this inlet is braceleted by the wide 12th Street vehicular bridge, a pedestrian bridge over the outlet channel, and the Lake Merritt Amphitheater. These human features now define the mouth of the lake.

The flow and ebb of tidewater past the scene is Lake Merritt's daily breath. Its windpipe, so to speak, is the narrow channel to the Bay where the water can be seen rushing, in or out, at almost any hour. With its tidal flow north and south and its vehicular flow east and west, Lake Merritt's mouth has the energy of a crossroads. People come to run, dream or drum. Birds come to perch, dabble, stalk or flock. There's usually a bay breeze here, and a wide sky. This place where land and sea and air converge is Oakland's natural center, its shimmering heart.

The Oakland Hills dominate the horizon at Lake Merritt's mouth, their serpentine skyline stretching from half-forested Grizzly Peak on the left to the residential gridiron of the Crestmont neighborhood perched high on the right. Several different bodies of rock make up the ridgeline, each in its own segment. A few more distant peaks, each made of hard lava, peer over the ridge: Vollmer Peak with its broadcast towers, the bald knob of Barberry Peak and tall, wooded Round Top. In front of the ridgeline is a nearer, shorter ridge, not always easy to see. It's a separate piece of the Earth's crust I call the Piedmont block. Between these ridges is the Hayward Fault, secret shaper of the whole. This book visits it all.

The subject at hand, though, is the landforms around Lake Merritt; we must lower our eyes and study the shoreline to see them. To visit them, an hour's walk is sufficient. To understand them, one must go beyond simple geography.

The geographer's approach to landforms is to describe and map and classify them. The geologist goes further to interpret them and learn their roles in a deeper history.

Downstream from the 12th Street bridge, the lake's outlet channel snakes in a narrow trough. On the channel's west bank, to the right, is the 1914-vintage civic auditorium next to the 1969-vintage Oakland Museum of California. Both buildings lie on landfill inside the original, natural channel shown on the old map. The original shoreline, the edge of the marsh, is visible farther to the right where 12th Street starts to rise near the museum's back end. Atop the rise is the natural platform of downtown Oakland, around ten meters in elevation; the county courthouse and neighboring landmarks stand on it.

The other side of the channel, on the left, is steeper and less obscured by buildings. Here the east bank rises to a platform of level ground, one that's a few meters lower—a difference big enough to measure, but not enough to notice.

These two platforms, so much alike, have completely different origins. The downtown terrace, Oakland's birthplace, is the subject of the next chapter. It is made of windblown sand. Here I'll focus on the other one, a marine terrace made of mixed sediment that accumulated under the sea. The terrace extends up the channel past the bridge toward the hills and sweeps around the shore.

Looking north over the lake from the amphitheater, we see the marine terrace on the right side, continuing nearly half a mile, past the white boat landing at the foot of East 18th Street, to the grassy rise of Pine Knoll Park. On the lake's north shore straight ahead, the terrace makes up all of Adams Point, the wide, wooded shelf in front of Grizzly Peak that houses Lakeside Park. Out of sight on the left, it also flanks the narrow north arm of the lake. Everywhere the terrace is present, its flat surface and its abrupt edge are plain to see on the streets.

This terrace is made of clay and sand and gravel, topped with clay. It dates from a time in the recent geological past that was quite different from today. The shelf's upper surface indicates where sea level once lay about 125,000 years ago. That was during an exceptionally warm break in the cycle of ice ages, when the global climate not only melted the great continental glaciers of North America and Eurasia but also shrank the ice sheets of Greenland and West Antarctica. The melted ice filled the sea some six meters above today's high-tide level.

Today's lake was then a wide, shallow bay. For perhaps a dozen thousand years, the Oakland streams flowed from the hills to this higher sea level and deposited sediment along the shore, where the tides swept it back and forth like a mason finishing a concrete sidewalk. Marsh vegetation took root and filtered from the estuary water a top coat of clay. The scene was lush; the low hills on the shore supported grazing and browsing animals that are extinct today. Then the glaciers reappeared, and gradually, as ice piled up on the continents, the sea shrank back and the terrace became land. Since then parts of the terrace have eroded, but much of it remains.

This information is a lot to recite about a simple shape on the ground. How do geologists know all that?

One might start with the sediment, exposed in a bluff near the tip of Adams Point where the shore has apparently never been modified. I've visited the bluff, with caution, during extreme low tides. It exposes gravelly clay, a hard mixture of soil and mostly pea-sized pebbles, none larger than a peach pit. The pebbles have somewhat rounded edges, lie in vaguely defined layers and consist of familiar local rock types from the hills. Evidently the sediment was polished in Glen Echo Creek on the way to a gentle shore, where it was washed in the surf for a while longer and then buried.

The elevated position of this material suggests two possibilities, both of them dramatic: either the land was low and then somehow rose, or the sea somehow was high and then fell. To clarify which is true, the geologist calls on a large body of global and regional knowledge.

This marine terrace has counterparts scattered farther north along the East Bay shore. At Lone Tree Point in the town of Rodeo, geologists collected fossil oyster shells from the base of the terrace and applied the uranium-lead dating method to determine their age: approximately 125,000 years. This date can be laid on the global timeline of the ice ages, a chronology established after decades of painstaking work. In that chronology, 125,000 years ago is during a warm period known as Marine Isotope Stage 5e or, more euphoniously, the Sangamon interglacial. High sea levels built marine terraces like this one at other places around the world. The chain of evidence is strong enough for me to accept and repeat this story, barring new evidence that might show otherwise.

Notice that I've interpreted a landform with a story that condenses a string of inferences based on geological evidence, some of it simple field observations and some of it exacting laboratory work. Several independent lines of evidence come together to confirm the scenario. The story is provisional, with room for improvement, but it stands as a nugget of fact that's true enough to be used elsewhere. That's how the science of geology works.

To put this age in perspective, recall that the Oakland Hills are roughly a million years old, a few million at most.

To me, with my head firmly in geologic time, that feels like last week. The age of the terrace feels like last night.

Beyond the marine terrace, the far end of the lake is framed by low hills, friendly ones studded with dwellings and shops, rising steeply from the shore a hundred feet and more. They're made of ancient river gravel and sand, cemented by mud and clay as firmly as sediment can be. The nearest bedrock is another mile farther away in the hills of Piedmont. These gravel hills and the more distant bedrock are two more puzzles, part of the million-year story told by the Hayward Fault. The lake, as the marine terrace has hinted, is a story told by the sea.

With that, let's look more closely at the lake. Again, the first thing is to separate the human imprint from the natural landscape. The city dealt with this large, unavoidable natural feature in its midst by changing everything about it. During its human history, the estuary has been an asset or a problem, depending on the purpose it was put to and whether one considers the water or the shore.

• • •

As a water feature, the slough was first a hunting ground and harbor for the Ohlones, who had a village at its eastern tip in Indian Gulch. During the Gold Rush, Americans came to the slough they sometimes called Laguna Peralta to hunt ducks. Access was easy enough for the Ohlones' watercraft of bundled reeds, or a settler's rowboat or small skiff, but only near high tide. At low tide the slough was a shallow channel surrounded by mudflats and marsh.

To the white people who settled this land, the slough was a barrier. The first Spanish expedition had to detour around it, irked by the steep hills and mosquito swarms at its eastern end. During the Mexican years, it slowed travel along the shore between the East Bay ranches. American settlers gathered in villages around two separate landings on opposite sides of the slough, San Antonio (soon renamed Brooklyn) on the east side in the 1840s and Oakland on the west side in 1850. Traffic between them was easiest by boat.

A few months after Oakland was incorporated in 1852, the town's infamous founder, the squatter Horace Carpentier —more about him later—built a bridge across the slough's wrist. It was a toll bridge, naturally, that the citizens complained about bitterly, sometimes drawing pistols rather than paying the agent. That was the ancestor of today's 12th Street bridge. More bridges were built nearby, for roads and for rail lines, and the lower half of the slough began to be filled in. Although both sides of Lake Merritt have been part of Oakland since the annexation of Brooklyn in 1872, it still marks the line dividing Oakland from East Oakland. Today, the interstate freeways bypass it on elevated roadbeds as if to avoid muddy feet.

In 1869 the mayor, Samuel Merritt, had a regulating dam built next to the bridge. The newly redefined lake immediately took his name. In 1870, Merritt pulled strings and had the state government quietly pass a law "to prevent the destruction of fish and game in and around the waters of Lake Merritt or Peralta, in the County of Alameda." The law banished the duck hunters while leaving the shore open to human encroachment. Soon gracious homes began to rise on lakeside lots carved like steaks from the holdings of Merritt and other prominent men.

Oakland didn't take Lake Merritt's wildlife seriously for many decades. Meanwhile, the growing town's untreated sewage fouled the water. An outlet tunnel built in 1876 carried the worst of it down 20th Street to the Bay. Around that time, the lake's floor was gaining an inch per year of filthy mud. In the 1880s, when the city's population more than doubled, sewage was linked to outbreaks of disease. In 1890, every doctor in Oakland signed a letter urging the city to drain the lake or fill it in. But the courts were still sorting out who owned the bottom of the lake.

When the litigation finished, in 1893, city leaders began to make Lake Merritt something more to their liking. Their projects changed the slough and its shore beyond recognition —except for the underlying geology. As the twentieth century began, a dredging campaign turned the muddy slough into the basin of open water we know today. Proper sewage treatment helped the water heal, starting in the 1940s. A pumping station in the outlet channel started regulating the lake's level, for flood control, in 1968.

None of these things were done for nature's sake. For a long time, the city's makeovers were human-centric. Firework displays were regular spectacles. Powerboat races in the so-called bird sanctuary began in 1927 and went on for fifty years. Old-timers remember them fondly. The lake served the city as an ornamental pool, just as the people and their leaders wished.

Then public attitudes toward wildlife began to change. More dredging in the 1980s helped remove decades of foul buildup. Aeration fountains were put in place, and public bonds funded massive renovations in the 2000s. The only boats allowed today are powered by sails, humans or electricity. Now the lake—the estuary—seems to be doing better. The water smells good, the birds thrive and we welcome our visiting guests from the sea. We know better how to work with the natural order.

The land around the lake, too, has been transformed in that time. In the 1900s decade, bond money bought out the lakeside estates, putting the entire shore in public hands, then the muddy margins were built up with dredging spoils. Along the eastern shore, a new street named Lake Shore Boulevard and a wide fringe of lawn were laid out on this landfill. Some spoils went into a little bird-nesting island in Lakeside Park, the first of what are now five. In the 1910s, the lake was lined with stone walls, backfilled with dredgings and broken rock from local quarries. Except for a small portion of Adams Point, the lake's entire shore was armored until the renovations of the 2000s opened a few spots for marsh plants to return.

Along the slough's outlet channel, landfill turned the wide, marshy passage into a tidal millrace. The newly made land was used to house a new campus for Laney College as well as the civic auditorium and the museum. In the 1950s, an amusement park beside the channel included a kid-sized railroad line and a paddlewheel boat.

At Adams Point, the marine terrace was made over repeatedly as the city evolved. In 1850 an early settler named Romby dug clay there in the oak woods and manufactured bricks, as others did elsewhere on the terrace. Then it became the property of Edson Adams, second of Oakland's three founding squatters. For a while it was a golf course.

The city acquired the Adams tract in 1907 to create Lakeside Park. Over the next fifty years, the flat ground filled up with attractions for every class of citizen—a beach, a boathouse, ball courts and a bandstand; gardens, a large marble fountain and an arboretum; a nature center, a duck sanctuary and a geodesic dome; a pony ride, an amusement park and a miniature railroad. The people loved them all.

My own favorite things in Lakeside Park include the landscaping boulders in the older parts. They come from all over the Oakland Hills, and they remind me of places up there I should visit again. Another thing I love, in the park's bonsai garden, is the display of suiseki—naturally sculpted stones, mounted and appreciated according to a Japanese artistic tradition. A well-chosen "mountain stone," a specimen the size of a bread loaf that evokes an alpine range, teases me with its conflation of physical scales and brings to mind the old Zen saying condensed in a song lyric: "First there is a mountain, then there is no mountain, then there is."

Elsewhere, landfill transformed the head of Lake Merritt, where a large colonnade and pergola was built in 1913. The structure opens toward the lake's outlet a mile away. East of the pergola toward the hills, landfill has turned the slough's marshy tip into the wide grassy flat of Splash Pad Park, flanked by hills of ancient gravel and crossed by Interstate 580 on its concrete legs.

Several streams once came together at this flat, and the eye can follow their valleys toward the hills of the Piedmont crustal block. Pleasant Valley Creek comes down Grand Avenue on the left, and Wildwood Creek descends Lakeshore Avenue, as it's called today, on the right, joining Indian Gulch Creek a block behind the freeway. Like other Oakland creeks, they now run in culverts beneath the roadway under round steel lids, although the heaviest rainfalls bring them erupting out.

In pre-settlement times, the streams were gradually filling the slough with mud and clay. Oakland sped things up by burying the creeks and finishing the job itself. Now the lake is firmly in human hands, but it was never static in nature's hands either.

• • •

Sedimentation is a global geological process that tends to erase lakes and estuaries. At the rate it's filling up, Lake Merritt would disappear geologically soon, in a few thousand years at most. The geologist is moved to ask questions that may sound odd. Why is Lake Merritt here? Why is it here now? Why is it unique in the East Bay? The lake and the ice-age landforms around it give us parts of the answers.

The pergola is a good spot to visualize this place at different times during the ice ages. Imagination informed by geology can bring us a fuller picture of the lake's history, embedded in a different kind of time, an elastic mix of then and now I call the geologic present.

In the human present we have a lake that we've re-made to our liking and can decide to change some more. In the geologic present, the deep past and present—the terrace and the shoreline—are superimposed. Geologists, knowing the lake's deeper history in the ice ages and the cycle of forces that have governed it, can see this place in all its states simultaneously. We see there is a lake, then there is no lake, then there is.

I've been referring to "the ice ages" rather than a single Ice Age. Since the beginning of the Pleistocene period, about two-and-a-half-million years ago, there have been some fifty ice-age cycles. Each cycle began with a steady worldwide cooling trend as large bodies of ice accumulated and spread, mostly on the northern continents. As each cold phase progressed, sea level fell by as much as about 130 meters, as deep as Oakland's tallest buildings are high. After a cold maximum, a rapid warming trend reached a peak as the ice melted back.

Unlike the daily ocean tides, caused by the gravity of the Moon and Sun acting on the Earth, the ice-age cycles are driven by oscillations of the Earth's orbit that bring it closer to or farther from the Sun at different times of year. The oscillations arise from the gravity of Jupiter, Saturn and other planets and are measured in tens and hundreds of thousands of years—rhythms gigantic and slow enough for the geologist to savor. These have happened throughout the Earth's history, but usually they haven't caused ice ages. That depends on the configuration of the continents, which slowly changes thanks to plate tectonics. Because the movements of the planets and the continents are precisely known, we can be confident that our current state since the Pleistocene, what geologists call an icehouse climate, will last for millions of years to come.

Right now, we're at a warm peak in the ice ages, the latest of many. It has lasted longer than any human records, so to anyone who's not a geologist it feels permanent. Nothing is permanent to a geologist.

During the previous warm peak of the Sangamon interglacial around 125,000 years ago, when the marine terrace was built, the place where the pergola stands was around six meters underwater. Seeing the terrace in person makes this easy to visualize. After that the latest ice-age cycle began, and it reached its cold peak about twenty thousand years ago, a time commonly called the Last Glacial Maximum. To me, living in the geologic present, that feels as recent as the dream I awoke from this morning.

During the Last Glacial Maximum, the Pacific coastline lay far out by the Farallon Islands, and all of San Francisco Bay was dry land. And here at the pergola there was no lake at all; instead the streams that flowed to this spot merged into what I call Merritt Creek. The creek cut deep into the land, no longer limited by sea level. In this freshly dug ravine, easily a hundred feet deep, the streambed probably had enough gravel for salmon to spawn, harassed by grizzly bears. Mastodons and giant ground sloths browsed on the slopes and drank from the rushing stream. Merritt Creek ran far below the level of the pergola where I stand. In the geologic present I see the pergola in a threefold state, at once here, six meters underwater and floating in the air.

As the Last Glacial Maximum ended, the sea rose back, the valley of Merritt Creek drowned and then filled to the brim with estuary mud. To find any traces of that buried ravine would involve drilling into the floor of Lake Merritt to take core samples, or conducting seismic-reflection surveys like the ones oil prospectors do.

Given all these facts, it's clear that the bed of Lake Merritt has been dug out and filled in, not once but many times. More than that: Lake Merritt is a rare state in the geologic present; usually it's dry land, a valley with a stream in it. This insight, sublime in itself, will have real significance later when we look at the flats and shoreline.

Standing on Lake Merritt's shore today, we can see both our role in the natural world and the natural world's role in ours. We aren't done with Lake Merritt; the landscape is not something we're ever finished with. Since 1850, we have turned a tidal slough into Oakland's cozy living room, centered around a great open-air aquarium. The extra global warming caused by two centuries of human greenhouse gas emissions is pushing the Earth into a climate state it hasn't seen since before the Pleistocene. We may have stepped beyond the geologic present, speeding up the pace of geology a hundredfold until it impacts the human present. Now the sea around the world is rising measurably, and within another human lifetime it will be a meter, maybe two, higher than today. As that time approaches, the city will need to remake our intimate civic partner.

Lake Merritt will need careful attention, followed by

thoughtful action. This will involve more than raising the bird islands and moving fences; for instance, a rising sea also lifts the water table in the adjacent land, outside the stone walls. We have options: we can enlarge the lake's outlet, raise the walls, sculpt the bottom, yield ground. Whatever we do will need to be coordinated with sea-level projects elsewhere on Oakland's waterfront and mesh with efforts by other cities all around the shores of the Bay. Civilization must grow smarter, wiser and more nimble as the water rises.

The sea is telling us something that the Hayward Fault denies. The fault repeats an old message with every earthquake: Earth has its own agenda that we cannot overcome, only adapt to. The sea, rising in response to human rather than cosmic causes, tells us a new one: our own agenda affects the Earth in ways *it* must adapt to. The entire planet is now our intimate partner. Slowly it's sinking in that we have powers we didn't know and obligations we need to recognize.

Oakland can set good examples based on this new, wider conscience. That would be a change, I admit, but Oakland from its earliest days has never shied from setting bold examples.

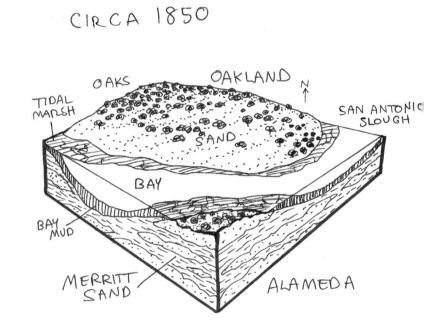

CIRCA 1850

OAKS   OAKLAND   N

TIDAL MARSH

SAN ANTONIC SLOUGH

SAND

BAY

BAY MUD

MERRITT SAND

ALAMEDA

DOWNTOWN

Look out any window in downtown Oakland, or around from any corner, and the land on all sides is flat. Oakland was born when three American squatters picked this level stretch by San Francisco Bay for a city-building scheme in the summer of 1850. It was a sandy plain thick with oak trees, well suited for a town site, and the fluid political climate favored men prepared to move fast and break things.

At the time, Don Vicente Peralta, whose land it was, dwelt a few miles away in a compound with his family, herds, and Ohlone ranch hands. He was one of a generation of Alta California ranchers at the tattered fringe of the Spanish empire and the succeeding Mexican republic. They had little ambition—or option, really—other than to pass the years on their range lands as part of a static neo-feudal aristocracy. But everything changed around Peralta and his peers in 1848, when the United States seized Alta California as a war prize. Two years later as the squatters launched their plot, the Gold Rush was in full swing and California was in the process of being admitted to the Union as its first Pacific state.

The founding of a great city on the east side of the Bay was long foreseen. Captain Frederick Beechey, visiting on behalf of the British Empire in the 1820s, was only the latest to say so as he assessed the broad cattle pastures along the Bay shore in the Mexican period: "In short, the only objects wanting to complete the interest of the scene are some useful establishments and comfortable residences on the grassy borders of the harbour, the absence of which creates an involuntary regret, that so fine a country, abounding in all that is essential to man, should be allowed to remain in such a state of neglect." In the 1840s, American visitors were aware of the same potential.

Then came three men in 1850, armed with gold, chutzpah and the US legal system, to make this oak grove into a new center of American civilization and grow rich with it. This wooded plain by the Bay, they could tell, had the bones of a proper city. And there was something special about it, something beautiful. Though they could see that the country differed from the rest of the East Bay, they couldn't tell how natural forces had made it that way.

They saw that unlike the shores elsewhere in the East Bay, which were either forbidding rocky slopes or soft marsh, here was a platform with the makings of a good harbor. Ships could reach it, with care, through a sheltered inlet and drop anchor by the sandy bluff along its southern edge, a setting made for wharves and piers. Unlike rough-and-tumble San Francisco across the Bay, it was land free of cold Pacific fog, unstable government and unruly culture, where streets could be laid and lots offered to homebuilders in an orderly way. Everywhere, mature trees provided ready-made shade for elegant estates. They saw land ripe for processing and sale.

The first thing needed was title to the land. The

squatters—ringleader Horace Carpentier, Edson Adams and Andrew Moon—erected a cabin and laid claim to three adjoining parcels of Peralta's property on the grounds that it was public land, the Mexican land grants to the ranchers being null and void. That was a lie; the grants were explicitly valid under the Treaty of Guadalupe Hidalgo, but the costs of proving so in the American courts bled Peralta to debt, and over the next decade he was obliged to bargain away most of his land.

The next thing needed was a government to administer the growth of a city. Carpentier used his influence in the new state legislature, where he had been appointed clerk of the Senate, to incorporate the town of Oakland in 1852—before the squatters had even gained title—and the chartered city of Oakland in 1854, where stuffed ballot boxes soon gave him the mayorship. It took two decades for the courts to untangle the ownership of the land. Then finally, in wave after wave, came the useful establishments and comfortable residences that Captain Beechey envisioned and Carpentier promoted.

• • •

This ground where the city was born is plain as a ballfield. It has a uniform topography but no unifying name. Early Oakland, let's call it, forms an oblong blob on the map extending east–west from the shore of Lake Merritt across West Oakland and north–south from West Grand Avenue to the waterfront. As it happens, the integrity of early Oakland is a geological integrity. The territory is an expanse of fine sand, without a pebble to be found in it, that stands between ten and forty feet in elevation and lies up to sixty feet thick. On all sides, the sandy platform slopes gently downward to its edge.

Professor Andrew Lawson of UC Berkeley investigated downtown's body of sand, marked its oblong boundary on the geologic map he published in 1914, and gave it the name Merritt Sand, "from its occurrence on Lake Merritt, in the city of Oakland." He thought it was a beach sand, but a few decades later its true story became clear. Several clues point to its origin: the frosted appearance of its grains under the microscope, its small and highly uniform grain size, the complete absence of coarse-grained material, the relative lack of bedding features. All are hallmarks of material transported and sorted by wind, what geologists call an eolian sediment. It is dune sand.

On the big scale, wind is a minor geologic agent compared to water and tectonics. Yet windblown sand has built Oakland's historical heart, just another special feature in the city's unusually complex geology. The dunefield of early Oakland is a legacy of glacial times.

A hundred thousand years ago during the cold phase of the last ice-age cycle, before the oak forest grew, before humans lived in the New World at all, California's environment was fierce and primal. As the high seas of the Sangamon interglacial receded, exposing the marine terrace around Lake Merritt and eventually leaving the Bay dry, the area became a cold, treeless grassland like the subarctic Aleutian Islands. Mastodons roamed the wide green coastal prairie, and what would become early Oakland was part of a windswept expanse of active dunes that reached for miles to the south and west.

The East Bay dunefields were born as the Pacific coast lay far to the west of today's beaches. The Golden Gate, newly opened by movements on the San Andreas Fault, was entirely dry except along its deep centerline, where a great river rushed brown with sediment from the continental interior. Inland, mountain glaciers were digging into the Sierra Nevada, carving out the valleys of Yosemite and Hetch Hetchy. They shed immense amounts of sand into Central Valley rivers that fast-tracked it through the Delta and spread it on the Pacific shore, where more sand washed down from the Coast Range.

One mild day in the Mojave Desert, I climbed the Kelso Dunes, a sprawling sixty-meter-high complex of excellent blond sand—hourglass sand, finer than salt. A few years later, a night-long dust storm in Death Valley that sifted kilos of grit into my tent showed me that to carry sand of this caliber takes winds of cheek-stinging force. The Merritt Sand is a little coarser than that Mojave dust, with heavier grains. To sweep up and sift and blow that sand across the Bay from the distant Pacific shore and pile it here to thicknesses of sixty feet—blanketing more than half of San Francisco with similar dunes along the way—those ice-age winds must have screamed like the sabertooth cat and howled like the dire wolf. On gentler days, clay dust mingled with the sand.

This era reached a cold climax and then began to ease. The ocean, rising as the glaciers melted, crept back into the Bay and up its sides. The Bay extended arms into the Merritt Sand that left three separate lobes standing above the tides: early Oakland, the peninsula of Alameda and little Bay Farm Island. The northern arm came up Merritt Creek, drowned its deep ice-age ravine and turned it back into a slough. People watched. Around ten thousand years ago, the sea stopped advancing.

By modern times, when written history began, the sand platforms of Oakland and Alameda were oak forests under Ohlone management. Later, after Spanish colonizers abducted the people and took their land, the northern arm got the name San Antonio Creek—the word *creek* having its old meaning of a tidal inlet with navigable water—and its extension was named San Antonio Slough. The creek and the slough flanked the Oakland dunefield on its south and east sides; on the west were coastal marshes and on the north a grassy plain.

This setting—platform, forest, navigable creek—was what drew the three American squatters in 1850. Unlike other locales on this coast, the flat was not crossed or drained by flood-prone streams. It was land ideal for realty and city-building.

City Hall, at Broadway and 14th Street, is a fitting spot to consider early Oakland. In a sense it is the city's origin point. Three city halls have stood here since 1869. It's the highest point in early Oakland at forty-two feet above sea level, defining Oakland's official elevation on every welcome sign at the city boundary. In front of the building, an old native oak spreads its limbs wide.

The grade from 14th and Broadway is barely perceptible, but in geologists' language it is quaquaversal—sloping downward in all directions. The slope is so even that a bowling ball flung hard down Broadway might roll nearly a mile to the water's edge at Jack London Square, where the city's first harbor was; launched up Broadway it would amble to a stop at 20th Street, where the road once crossed a small stream valley. The gentle, even grade was singled out for praise by Horace Carpentier and his fellow boosters. The firm sandy ground was naturally made for drainage and gutters and roadbeds and sewerage to be put in place with minimal cost.

The three squatters hired a Swiss surveyor named Julius Kellersberger to lay out a town on their blank slate of land. He started by drawing a straight line northeast from the harbor to this low rise, where it met the main road up the East Bay to the northern ranches, now called San Pablo Avenue. Kellersberger's line became Main Street, soon re-named Broadway. He made this principal avenue extra wide at 110 feet. Oakland has done big things from the start.

On Kellersberger's plan, Broadway was the spine of a gridiron of parallel streets, seven on each side. Crossing them at strict right angles, a set of streets numbered First to Fourteenth marched up from the waterfront.

That orthogonal grid was Carpentier's fixation, his ideal city plan. The grid was a touchstone American pattern replicated across the Great Plains on government-surveyed lands, divided into townships exactly six miles square. Among its other real and imagined virtues, gridded land is easy to subdivide and sell. The town grid was an emblem of the Westward Expansion, a perfect seed crystal set upon lands that history had conveniently made vacant for their new white lords. Carpentier had in mind a suburban city of wealthy men like himself, with room for all their fine fenced estates on a peaceful tree-shaded plain.

The vision of the grid roused Mayor Carpentier to lawyerly eloquence. "Acute and obtuse angles in a city are opposed alike to beauty and convenience," he argued in ve-toing a street ordinance. He lauded "the magnificent vistas which a system of straight and rectangular streets would afford, looking out in every direction upon the craters of the bay or the mountains." And it was true—the streets by lucky chance lined up with worthy landmarks we enjoy to this day: down the numbered streets rise Mount Tamalpais to the northwest and Fairmont Ridge to the southeast, and the named streets frame views of Grizzly and Vollmer Peaks toward the hills and San Bruno Mountain across the Bay.

There is something mythically satisfying about those cardinal points, and while Carpentier was rightly despised as an "unscrupulous grabber," that happy coincidence, that geographic *feng shui*, seemed to tickle the cold-eyed speculator's heart. He saw a here here. And although the city did grow acutely and obtusely like an oak from an acorn, gaining complexity with each decade and annexation, his grid remains.

• • •

The Spanish colonizers opened the written record of this part of the world in the 1770s, describing it as a place where evergreen oaks—*encina* in Spanish—formed a large, well-tended forest they later named the Encinal de Temescal. The encinal directly inspired the name of Oakland.

Our local coast live oak, *Quercus agrifolia*, produces abundant acorns, and the Ohlones relied on the encinal for that staple food and trade good as well as firewood. Because the encinal had no running streams, it was a place for work camps, unlike the Ohlone home village at the head of San Antonio Slough. The tribe set gentle, timely burns to clear the grove of dead leaves and underbrush, driving off pests with cleansing smoke and restoring soil nutrients with ash.

Although the Americans in early San Francisco enjoyed excursions to this great grove on the *contra costa*, the other coast, by 1850 the encinal was not what it once had been. With the Ohlones sixty years gone, the trees were old and their ranks thick with underbrush. None of that mattered to the occupying Americans, who cut down the oaks to feed the firewood trade. Wood and charcoal fueled everything at the time, even locomotives and factories.

The Encinal de Temescal may have been the largest stand of coast live oaks in existence, just as the redwoods of the Oakland Hills were once the greatest of their kind. Mayor Carpentier, who gave Oakland its name, called the oaks "the chief ornament and attraction of this city" in his in-augural address and strongly urged that they be protected. Leases forbade cutting them down. But the native trees did not last. In 1868 Titus Fay Cronise, in his inventory of California's natural resources, described Oakland as sitting within "what was once a fine grove of 1,500 acres of ever-green oaks," and contemporary photographs show only scat-tered specimens on some of downtown's unpaved streets.

One of these aboriginal trees, a "gnarled veteran" in front of City Hall, survived a hundred years into the 1910s. A young oak from Mosswood Park took its place in 1917, ded-icated to the memory of the Oakland author Jack London. That's the one at City Hall today, gracefully aging under an arborist's care.

Two other ornaments and attractions of the young city lay beneath the encinal. These, virgin soil and fresh ground-water, were its first geological assets.

Downtown Oakland's soil was a sandy loam that when treated well was prodigiously fertile. This was no real-estate legend invented by promoters, though promote it they did with stories like this from the early days: "On the block bounded by Market, West, Fourth, and Fifth Streets, Mar-shall Curtis raised carrots that pulled the scales down to the 18-pound mark." Bounty as absurd as that carrot, and the 81-pound squash and the 200-pound sugar beet, came from a super-surcharge of organic matter and natural nitrogen compounds that had built up for centuries in the encinal's topsoil. Natural land is this way; it's why the American des-ert always bloomed when farmers added water. Plantings in virgin soil tend to burst into spectacular growth, and unless these nutrients are diligently replaced, they're eventually used up and the soil is depleted. The problem is as old as agriculture.

Early histories boast of phenomenal farm yields up and down the East Bay. Virgin soil made Alameda County the wealthiest in California for many years. Naturally, the resource was called "inexhaustible." But exploitation to exhaustion is what Americans have always done. Once the soil was depleted or paved over, farming shifted to the Central Valley, where today the agricultural industry feasts on the world's largest area of Class I soil and water from the shrinking snows of the Sierra Nevada.

After its trees and its soil, decent well water was the third of early Oakland's ornaments and attractions, available at every lot in veins of coarse sand and gravel no more than sixty feet down. Naturally, the early boosters called it "inexhaustible." Many of the first wave of settlers were men with Gold Rush fortunes who surrounded themselves with landscaping, as luxurious as possible, and relied on wells and windmills to water their estates. A shallow layer of hardpan purported to shield the water-producing layers, or aquifers, from surface pollution; not for long. Most downtown wells were abandoned within thirty years as the aquifers were exhausted or tainted and as water companies came into being.

Local water still serves the last piece left of Carpentier's residential vision in the Kellersberger grid: the modest half-block estate of the Pardee family, between 11th, 12th and Castro Streets. Mature oaks and lush plantings surround the two-story main house, built in 1868 and now a museum. Behind it stands the old water tower, a lighthouse-like structure with a flat top where a tank once supplied pressure for the estate's plumbing. A whirling windmill pumped the tank full from a well in the tower. All of Oakland's first homes used well water—hard water, but sweet and safe for all purposes. The Pardee House well still keeps the grounds green even in drought.

The estate has other connections with Oakland's geology. George Pardee served as mayor from 1893 to 1895 and governor of California for a four-year term starting in 1903. When the San Francisco earthquake struck in 1906, Pardee took the train home from Sacramento to oversee the state's response. And in 1930 he became the first president of the East Bay Municipal Utility District, the agency that finally gave Oakland a water supply worthy of a great city.

Right across Castro Street from the Pardee estate now lies a great trench in the sandy platform where a multilane freeway, Interstate 980, roars with high-speed traffic. Both 11th and 12th Streets span this chasm and offer views of the hills above and the tumult below.

Whenever I make this crossing, bathed in the freeway's noise, I'm reminded that petroleum—oil and gas—in roughly two life spans has deeply transformed American society. The freeway, with its gasoline-powered cacophony, is one handy example. Today the petroleum age is starting to decline.

Scientific ingenuity and the wholesale consumption of petroleum and coal brought us literally superhuman powers. Humanity jumped its old tracks and became something new under heaven. Stable times, marked by age-old cycles of repeating seasons and little-changing lives, gave way to a long surge of progress without precedent, a technological tsunami fueled by deposits of ancient carbon, that has so changed the Earth's environment as to threaten its fitness for civilization. For our own good, we must rebottle the genie and bury it again.

Geologists helped bring about the petroleum age. They were complicit. It's one thing to see the Earth with the eyes of geologists—a puzzle, a comfort, a pleasure, a refuge— and another to recognize the crucial role of geologists in making and supporting the modern world. Today, geoscience professionals have a wide choice of essential careers, maintaining what we have and helping the world move forward. They're ready to help whoever is ready to listen.

Just a lifespan from now, oil and gas will not be pumped from the ground but be manufactured, in small quantities, for niche purposes and historical reenactments. The electric freeway below me will be quieter than today, and it will smell better. Planners have schemes that may turn this trench into a rail-transit tunnel with a surface boulevard on top. If that comes to pass, visitors to Pardee House will hear the birds again.

. . .

The downtown soil, though it's no longer farmed, is still a city asset. The stuff beneath the topsoil is firm, deep and predictable; it serves well the buildings that rely on it. I see it in every foundation hole, fine-grained sand infused with clay, a clean material the color of caramel. Plugs of subsoil fresh from the core drill are dense in the hand, like chilled cookie dough but far stiffer. We're lucky to have this good foundation material, because bedrock is out of reach in the East Bay flats, buried by hundreds of meters of sediment. Downtown Oakland has grown upon this base into a new encinal of concrete, stone, tile and steel, with a new generation of towers clad in glass that suit this young century in the ways they play with mass and light. Yet a stroll around the Kellersberger grid passes older parts of early Oakland with connections to its geology.

On Broadway between 10th and 8th Streets is a historic district called Old Oakland, preserving several intact blocks of what was the business district in the 1870s. The new transcontinental railroad, completed in 1869, ran down 7th Street and brought with it a burst of prosperity reflected in the buildings' ornate fronts and solid construction.

These buildings were strongly influenced by the area's geology, in a way that had nothing to do with their foundations. The major Bay Area earthquakes of 1865 and 1868 led to California's first earthquake-resistant architecture and an urban fabric that visitors noted: buildings were very solidly made, and they never rose higher than five stories. By the time of the great 1906 earthquake, local architects had advanced their art, and in San Francisco tall structures like the Palace Hotel, tied with iron bands and designed for lateral loads usually assigned to hurricanes, survived the shaking well. Instead they succumbed to fire. Oakland luckily avoided the fire that time, and many buildings of the era survived.

Speaking of fine old buildings, downtown is an archive of a century's architectural fashions in stone that invite the eye and please the hand. Geologists, naturally, keep an eye on the stone used in buildings wherever they go. They linger in lobbies and hang out on corners, trying to avoid suspicion as they finger a particularly nice mineral texture or stray fossil.

Stone has gone in and out of style for building exteriors, alternating with timber, brick, stucco, steel, concrete and tile. Fans of those worthy materials—especially tile—have their own set of treats in store as they explore Oakland's downtown.

Oakland City Hall, dedicated in 1914, is a remarkable period piece: a stack of offices made up as a wedding cake. Its design was called a "monument to metropolitan progress and energy" and "the most perfect example of modern municipal building to be found in the United States." Its walls are faced with off-white Sierra Nevada granite and lavish terra-cotta decoration. The interior is an inspiring blend of marble, tile, brass and dark wood.

Like all the building stone downtown, the "Sierra White" granite is worth a closer look. It's a fine-grained rock, more salt than pepper, from the Raymond quarry in Madera County. Its lines and edges are as crisp as the day they were cut. This granite was lauded at the time as "a celebrated California building material which lends distinction to the finest structures on the Pacific coast." It sent a message of stability, nobility and strength. The city preferred it over sandstone facing for the sake of beauty and pride, despite its greater expense. Many California buildings of the era made use of this excellent granite. Sierra Nevada granite has another role in this book as a key part of Oakland's deep geologic history.

Surrounding City Hall, what looks like a granite-lined ditch or a dry moat is actually a clear space that allows the whole structure to shimmy by a foot and a half when a major earthquake shakes it. The foundation was rebuilt after the 1989 Loma Prieta earthquake, and today the building rides on more than a hundred blocks of rubber the size of cafe tables. Oakland City Hall is a trend-setting case study in how seismic retrofitting can save valuable buildings and help cities adapt to their inherent geological hazards.

The sandy plain has many more notable showcases of stone. At 20th and Broadway is the I. Magnin building, an Art Deco structure built in 1930 that once housed a department store of cherished memory to East Bay grandmothers. Its whole ground floor is extravagantly faced with "verd antique," a deep jade-green rock slashed in all directions with white veins. It comes from a single quarry in Rochester, Vermont. Stone dealers market it as a variety of marble, but in fact it's something more exotic: serpentine rock. This stone was once in fashion for trim and accents, seen in dozens of downtown Oakland buildings large and small, and it still appears in newer structures, as distinctive as ever. Oakland has its own large body of serpentine rock in the high hills, but it's not of commercial quality.

Two other interesting buildings are at 20th and Harrison Streets right at the lake. The 1960-vintage Kaiser Center complex has large, flat walls textured with fist-sized rough pieces of coarse-grained white marble—again, not ordinary marble but harder, more durable dolomite rock from the Natividad Quarry south of the Bay Area near Salinas. Dolomite, too, can be found in the Oakland Hills. And across the way is the statuesque Lake Merritt Plaza building, circa 1985, a curvy high-rise clad from top to bottom in polished "Texas Pink" granite, the brand name for stone from a quarry near Austin. Its pinkish color comes from the mineral feldspar that pervades the stone in ragged crystals the size of golf balls. I think it's the most beautiful building stone in Oakland.

Several historic downtown churches are faced with rugged blocks of blond sandstone from Bay Area sources. To my knowledge, only the First Unitarian Church, at 14th and Castro Streets, used a truly local sandstone, from the small McAdam quarry in the high hills just over the city line in what is now Redwood Regional Park. Oakland rock is so heavily fractured, thanks to the Hayward Fault, that among other "vexatious difficulties" of that 1890s construction project, "quarrymen were unable to deliver this material in sufficient quantities."

At the time I write this, the fashion is polished limestone with muted colors and finely layered texture. My favorite recent example is the Dellums Federal Building at 13th and

Clay Streets, with a cladding of blond limestone, accents of verd antique and pavers of dark granite. In the echoing glass atrium between its twin towers, a great mosaic floor depicts a map of the San Francisco Bay Area in a rainbow of natural stone colors.

• • •

Most cities owe their germination to an accident of geology; Oakland happened to sprout on a field of sand dunes, its youngest landform. Deep time goes much farther back in Oakland's other places. Keeping the preliminaries I've laid out in mind—the forces of the fault, the cycles of the sea and the impulses of city-builders—let's pay them a visit.

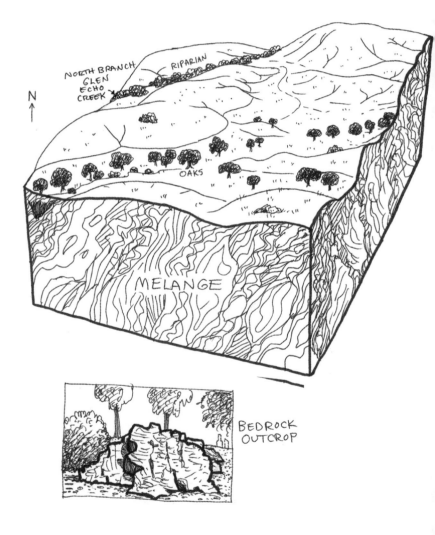

NORTH BRANCH GLEN ECHO CREEK
RIPARIAN
N
OAKS
MELANGE
BEDROCK OUTCROP

MOUNTAIN VIEW CEMETERY

It takes an odd mind to envision a place like this, to behold a rough tract of outlying rangeland and think, *Here I shall build an elegant burying ground befitting a great city.* It takes an enterprising mind to bring such a scheme to life. Investors must be gathered, an operating plan devised, a perpetual endowment arranged to support a business project based on the monetization of solace and a couple hundred acres of beautiful terrain. The result, a century-and-a-half later, is the Bay Area's finest landscape.

Mountain View Cemetery is private land but a de facto park, draped on a picturesque slope, graced with lawns and specimen trees and sprinkled generously with monuments of all kinds. The hillside nestles it in a topographic bowl like a pair of cupped hands, the upper watershed of Glen Echo Creek. With my own odd mind, I visit the cemetery to lean and loaf at my ease, observing the landforms under its manicured surface and paying its rocks respects.

The cemetery is a geologically interesting place. The hills, by my estimate, are about a million years old, the rocks in them something like eighty million. But here human time overlays geologic time; dates are displayed wherever I look. Monuments of artisanal stone mingle with outcrops of natural rock. And the cemetery's people keep joining my thoughts. Although most residents in Oakland's hills keep to themselves, in Mountain View they crowd my eyes, showing me their names, and plead for recognition. I've come to recognize many, some of whom were geologists themselves. Between the rocks and the departed who rest among them, my reveries can get rather intricate.

Samuel Merritt (1822–1890), of Lake Merritt fame and odd mind, arranged the syndicate that gave the cemetery its charter in 1863, and Frederick Law Olmsted (1822–1903), famed for his role in designing New York's Central Park, gave Mountain View its soul. Like all of the hills at the time, the area was grasslands, green and gold in season. On this canvas, Olmsted laid out the core of the grounds in 1864 on his novel principles, seeking resonance between the pleasing organic contours of well-tended land and our relationships with the departed. His plan well suited this natural amphitheater with its rocks and knolls and vales.

The site is not in the high Oakland Hills but in the more modest heights of the Piedmont block, west of the Hayward Fault. Mountain View preserves a prime part of this block as an island of open land. Its underlying form is intact despite being prettified and covered with graves. More open land borders the cemetery's north side in the form of two smaller cemeteries and a golf course beyond—another kind of costumed countryside en-balmed in springtime green.

The cemetery's grand entrance is on flat ground at the top of Piedmont Avenue. This area, the uppermost floodplain of Glen Echo Creek, has been built up a bit and the creek buried, to put any thoughts of flooding out of mind.

From the circle by the entrance one may gravitate up slopes in any direction. On the left is the compact Jewish cemetery, where visitors place pebbles on the gravestones. In the middle is the one straight roadway in Olmsted's design, a long allee that forms the axis of the cemetery plan. The first of four fountains along the way was contributed by Anthony Chabot (1813–1888), the "water king" buried in plot 9. Rightward is how I like to go, around the mausoleum complex, then across the high ground and back down on the left in a widdershins loop.

The cemetery reminds me that geologists are humans first. They secure grave plots under contracts pledging to maintain them for time without legal end—even though geologists know that within deep time, landscapes are as melting ice, and even though they may admit that human perpetuity is of no more moment than the swash of a wave at the beach.

Mountain View's monuments are a cross section of historical fashions in graveware. Near the tall pyramid of Senator William Gwin (1805–1885) is a chip of unworked granite the size of an armchair, inscribed with the name and dates of Joseph Le Conte (1823–1901). He was a prominent geologist, someone who left a legacy. As a founding professor at UC Berkeley in 1869, Le Conte helped make it the nation's preeminent public university. He taught geology there for decades. His college textbook went through five editions. With John Muir (1838–1914) and others he founded the Sierra Club. When his heart gave out during a visit to Yosemite Valley, friends purloined this boulder from Glacier Point for his headstone.

Joseph Le Conte was one of my tribe. He published papers on the volcanic rocks of the West and the plutonic granites of the Sierra Nevada. He was a bearded and courtly scientific aristocrat who held high offices in the professional societies. Every age reconsiders its heroes once they fade from living memory, and lately he has been found wanting.

His geological theories are obsolete. My tribe doesn't hold that against him. We know that our science is young and the way filled with uncertainty and error. The point is to keep the community in vigorous ferment, and Le Conte did that. And the man loved rocks. Unfortunately for scientific posterity, he advocated a cooling, shrinking Earth. For a few decades this was the preferred narrative of how seas might sink and continents with their mountain ranges rise, like the wrinkling skin of a drying apple. As facts accumulated and thinking deepened, the shrinking Earth was abandoned for narratives more nearly true.

We swim in the intellectual currents of our times, and those have moral correlates. In Le Conte's time it was widely presumed that nature progresses, and he conceived that Darwinian evolutionary theory not only explained the origin of species and descent of Man—and human supremacy, of course—but also applied to natural progress in other realms. "There are certain laws underlying all development," he wrote in his *Elements of Geology*, "certain general principles common to all history, whether of the individual, the race, or the earth." That may have impressed his freshmen readers, but no scientist of any kind would assert such a thing today; it's just not true. Today we hold that the Earth follows no general pattern of development, a concept that presupposes a plan, whereas science does progress, in the direction of truth.

Worse, Le Conte relied on this simplistic Darwinism, as many did, to support white supremacy. An unrepentant racist born in Georgia to a slaveholding family, he held that humanity was divided into a hierarchy of races, among which his own race was further *developed* than others. And he defended this harmful view his whole life. Posterity has the right to revise its opinion in light of new times, and Le Conte's grave gives me pause.

Up the hill from Le Conte's dusty boulder are some of the cemetery's oldest plots. Here the best stone that quarries can provide has begun to show signs of weathering. Marble monuments are sugaring, no longer polished and creamy, after a century's exposure. Granite markers weaken and stain as rain and air attack their dark minerals. Slate and sandstone soften their crisp lines with age. Not even solid rock stays pristine for long.

At the hill's top by the boundary wall is a curiosity, the plot for the Benevolent Order of Elks, where a bronze elk stands atop a rustic construction of blue-green stones. These are serpentine rock, probably from a source in the Oakland Hills.

From here I tend to wander along the cemetery's edge, up the flank of Glen Echo Creek's headwaters. A hundred yards or so ahead are three unexpected ponds within a fringe of trees. These reservoirs were dug and dammed in the 1880s to water the grounds and help keep the floodplain below flood-free. Their water bathes the grass, percolates back into the creek and enters the Bay at Lake Merritt. In season, there are ducks.

Past the ponds is plot 1, where two of the occupants have an earthquake connection. Joseph Josselyn (1837–1868), the promising young earthquake victim, is here under a slim sandstone obelisk. George Pardee (1857–1941), the California governor who dealt with the 1906 earthquake, is in the family plot near a pall-draped obelisk of white marble. And here too is Joseph Emery (1820–1909), whose first fortune came from rock quarries, but his grave has no marker.

He founded the flat, stoneless little city of Emeryville, on the Bay between Oakland and Berkeley.

The land the cemetery sits on could have become a rock quarry itself. Several former quarries lie around its edges. Indeed, Mountain View has quarried rock from its own hilltops as new sections of land were prepared for new plotholders. The rock is why I like these hills so much. It's shaggy, indigenous rock that complements the clean-cut monuments.

In plot 6 a low rectangle of polished coarse-grained granite bears the name of Nicholas Taliaferro (1890–1961), another prominent UC Berkeley geologist. He was not a grandiloquent philosophizer like Le Conte. He was hard core, a short, mustachioed guy nicknamed "Tucky," whose colleagues found him "gruff at times, and his vocabulary could be salty." He posed challenges to his freshmen, starting with his name, one of Italian origin but pronounced "Tollifer" in the old Kentucky way. He would bring classes to the hills above campus, seat himself by an outcrop and have a smoke, not saying a word. The students would hesitate, huddle, scatter and start taking notes on anything they could see. After a while he'd find another spot, sit, smoke, and so the afternoon would go.

When Taliaferro wasn't teaching, he was in the field mapping rocks. It was estimated that he walked some fifty thousand miles in the Coast Range and the Sierra foothills. He would have done more had he not been killed by a wayward car in downtown Lafayette.

He also spent part of his career with oil companies, helping them extract the ancient carbon that burdened our skies with heat-holding pollution and fueled the car that took his life. My tribe doesn't hold that against him— geology has always been civilization's handmaid—but we do recognize irony. We are only as wise as we are today.

Taliaferro has two legacies: one is that every serious paper about the central Coast Range cites his work. The other is that in running UC Berkeley's annual Geology Field Camp for thirty-three consecutive summers, he personally trained hundreds of the profession's eagerest students. In a field where mentors and personal interaction are important to a science-based career, he was hugely influential.

I find it fitting that Taliaferro lies near the wildest part of the cemetery, the part too steep to ever hold graves. This tree-studded grassland beckons me off the paths, up the slope. The views grow wider as I climb. Overhead an occasional hawk hunts voles. Burrs clutch my feet, and chiggers nip me if I sit too long. For a while, among the scattered oaks and resilient poppies, I can imagine that Oakland's original wildland felt like this.

And here are the native rocks: humps the size of pianos or cars, seeming to grow from the soil. Unlike the finished grave markers or the tumbled cobbles of streambeds and beaches, they're covered with lichens and mosses. I must find tiny gaps where the stone is visible before I can name their rock type.

• • •

Asking a geologist "What are rocks?" is like asking a chef what food is, or a writer what words are. They will smile, pause, and answer, "The real question is what rocks mean." Rocks are more than lumps of mineral matter. Rocks are things that have happened, results of particular events. Every body of rock was made in a specific place and time, and it's stamped with traces of that environment as surely as a serial number, if we have the skills to read it—that belief is what drives geologists.

To learn to read those traces, we study rocks where

they're forming today. At active volcanoes, we fact-check ancient lava flows against new ones. Far below the seafloor where mud is being compressed into mudstone, drillships retrieve core samples. In the laboratory, we put rocks under high pressure and heat, taking them briefly to the truly deep Earth where their minerals turn into those found in metamorphic rocks. After enough of this work, geologists can go to the rocks and seek answers to the question, "What kind of world made you?"

Every city has its own geological basement, full of dusty records made of sediment and stone. Every place has samples of the past worlds that once were there. Some of those past worlds—say, Nebraska in the Cretaceous period —were gentle places, far from the tectonic conflict zones of their time, but even their rocks register faint signals from distant rifts and ranges. California, though, has always been a geologically busy place, and Oakland's rocks have always been near the action. The ice-age changes and the vigorous tectonics of the Hayward Fault that I've already mentioned are only the most recent parts of a longer history.

A cemetery pulls one's attention into the ground, though not everyone notices the same things. The natural soil is not deep in much of Mountain View, but beneath it the rock is easily worked, mostly shale and sandstone in thin, brittle layers. The shale is a creamy-looking claystone and the sandstone is visibly gritty; both can be seen best where roadways bite into a hillside.

Looking around, one may notice that the cemetery's landscape is not like most of Oakland's rocky hills. There's something soft and irregular about it, without a strong underlying structure. The landscape doesn't show many bones. The thin-bedded strata don't stand out in ridges but weather into smoothly rounded shapes, sprinkled with the little outcrops I mentioned. These lumps of hard contrasting stone, like raisins in a cake, are too large to ignore but too small to map. The whole ensemble, raisins in cake, blocks in matrix, is a composite called melange.

Melange forms a peculiar terrain that's found throughout the Coast Range: lumpy, rounded, strewn with blocks. I find it familiar and endearing. Geologists have mapped it as part of the Franciscan Complex, named after San Francisco where melange underlies much of the urban carpet. Seeing melange here outside its main habitat, groomed and ornamented with trees, can bemuse geologists who know it from genuinely wild places. There's another large area of melange at the other end of Oakland in undeveloped Knowland Park, in the hills above the zoo, where some of the blocks are the size of houses. To me that place feels more like it should.

Geologists have a complex relationship with wildland, as most Americans do, but with a few extra wrinkles. Some are inherent in the profession and some are inherited, a cavalier legacy from the nineteenth century. And here I think again of Joseph Le Conte.

In founding the Sierra Club, Le Conte and his allies cultivated an aristocrat's romantic vision of wilderness. Although their mission was to access, cherish and preserve wildland, their vision was not about preserving the Sierra Nevada for its primeval residents, nor about accessing it to make a living there, but about cherishing the land as an inspiring setting, best suited for quiet enjoyment in fair weather. Early Sierra Club outings treated the mountains as a stage for alfresco amusements, an exclusive rustic resort where the firewood was free. The Sierra was their club; the members had no business in the mountains, although Le Conte did have his science.

Le Conte first visited the Sierra in 1870, and his journal

from that trip to Yosemite mentioned the Indigenous people only to complain of their thieving habits, the way the tourists speak today about the bears. But their world was vanishing fast. Soon enough, visitors placed their marks upon their newfound stage the California way. Once the Indigenous inhabitants had been evicted under national authority to establish "forever wild" national parks, recreational climbers showed up to erect stone cairns, containing notebooks for visitors to record their conquest, on peaks whose sacred names had been erased.

No question, the Sierra Nevada is still a wonderful place, one the whole world cherishes. And many visitors assure themselves that wilderness as God made it seems more noble, more true somehow than inhabited land where people forage and hunt and set fires. But the High Sierra we see today was not created by a wave of God's hand. Before the Spanish explorers named it for its permanent snows, the Sierra Nevada was a place the California tribes had managed for centuries, a place cherished because it was lived in. The mountain groves that captivated Le Conte were not primeval woods but a former human habitat, vacated at gunpoint and set adrift. Without tribal management the landscape has changed—still sublime, but not the same, a place everyone visits and few truly inhabit. And before that, the Sierra evolved on its own for over a hundred million years, a geologic history entwined with Oakland's deep history. My tribe finds that deeper history, written in rocks and landforms, compelling whatever costume the landscape happens to wear.

In Oakland, too, the land was not wilderness but a place where the Ohlones groomed their hills as grassland, rich in useful plants and vivid with native flowers. After the tribes were driven off their land, within decades the hills became prone to wildfire and the woods grew rank. Mountain View Cemetery takes good care of its grounds, though. Sometimes I see the Ohlones' lost meadow mimicked here in Olmsted's lawn sprinkled with feverfew. And the underlying ground, what I come here to see, is so nicely visible without the dwellings and fences of landowners.

Land rid of people has an extra appeal to geologists. The great quest of all scientists is to understand the universe on its own terms, transcending human lives and human senses. But remote study sites where people are few and geology dominates are motivating beyond their scientific relevance. There's something pure about their beauty. A strange attraction of space missions to other planets, noted by many geologists, is the uncanny images of alien land untouched by life itself—true and perfect wilderness.

My tribe's attitude toward wildland mixes insight and possessiveness. While we see the landscape like everyone else, we're also attuned to the space beneath it, and we don't stop with the view. With our eyes, we eat with relish the inner organs of hillsides. Landforms hint to us of larger structures and deeper histories within them. With training and experience, it comes at a glance.

My tribe can find it hard to leave wildland alone, and sometimes our behavior is troublesome. I've watched geologists pilfer stones and hammer on outcrops inside parks and military reservations without getting the necessary permissions, privileged by their curiosity. I've found their core holes drilled in scenic outcrops, and I've seen well-regarded researchers be censured for mistreating ancient pictographs. I wouldn't do such things, and professional codes of conduct as well as laws proscribe them, yet I too have climbed gates and crossed fences.

But I also treasure memories of group visits to special

lands, thanking the owner with a round of applause. And I recall when the late Eldridge Moores (1938–2018) bade me step back from a delicate feature in a roadcut when I sought to brush it clean. He said that disturbances by casual visitors would add up with repetition. Though it had no sign, the roadcut was precious. I remember his lesson each time I visit Oakland's most special outcrops. I would no more hammer at those rocks than I would chip at one of these gravestones. And now I rarely collect rocks.

In sum, the geologist's eye is a bit subversive; it ignores boundaries and other restrictions, legal or sacred. With equal curiosity we scrutinize the stones of churches, the floors of civic buildings and the walls of fancy lavatories. And so it is with me and this cemetery. It's a handy example of what California geologists, with easy familiarity, call Franciscan terrain.

. . .

This body of melange continues beyond the cemetery through much of the ridge west of the Hayward Fault that holds the town of Piedmont. This crustal block is an isolated piece of the Franciscan Complex, a cryptic mixture of rocks that makes up much of the Coast Range and reappears in later chapters. The person who named the Franciscan, in 1895, was Andrew Lawson (1861–1952), the link from Le Conte to Taliaferro in UC Berkeley's dynasty of renowned geologists. He too was buried in Mountain View, but was later removed to a family plot in Toronto.

The people whose graves I've mentioned had long lives. My tribe considers that typical. The outdoor exercise is part of it, but another thing may be that progress in geology is typically gradual. It pays to stick around. Advances are tentative, and firm consensus is hard won. A science rich in particulars and still weak in generality, geology needs large bodies of raw knowledge. And there is no end to rocks.

The saying goes that the best geologist is the one who's seen the most rocks. More than that, one must *interrogate* those rocks, and think well beyond the rocks themselves. I think brain work like this keeps geologists vital well beyond middle age. And certainly land and rocks are beautiful. Nicolaus Steno (1638–1686), geology's founder and literally its patron saint, distinguished three grades of beauty: beautiful are the things we perceive, more beautiful are the things we understand and most beautiful by far are the things we do not yet comprehend. Geology keeps me happy as the years pass.

At the highest back end of the cemetery are wooded and meadowed hillslopes full of bedrock blocks and signs of deer. Much of this land will probably be wild in perpetuity. Here the seasons and the native oaks hold sway.

Many of the best bedrock blocks are scattered on the slopes above Millionaire's Row, where the grave markers lie at ground level for easy mowing. Whichever way I take across the top of the cemetery, I'm sure to encounter a few. Some blocks consist of greenstone—black lava metamorphosed to a greenish gray and riddled with white veins. Others are rugged bodies of dark chert, a hard stone with no visible grains and a waxy, lustrous look. Franciscan chert can be as red as fresh beefsteak or, less often, a thrilling olive green. It often has regular layers a few inches thick, contorted in intestinal shapes. And it can also be marbled, like brisket, with white hydrothermal quartz. A hand lens will sometimes reveal the shadowy skeletons of one-celled protozoa called radiolarians within the chert, a cemetery's worth in a single pebble.

Each block is individual, none inscribed with a name or date, each with an intricate history that has brought it here. Each one warms in the sun; some invite a leisurely sit to take in the wide, harmonious vista. I judge it the Bay Area's best view, with delights in every direction. The cemetery's genius is this scene, a blend of high and low, of distant "borrowed scenery" and intimate framed vistas, groomed for mutual admiration, encompassing central Oakland, San Francisco, the Bay Area and the great flat Pacific outside the Golden Gate.

I said earlier that although the Earth is not directed and does not *develop*, science does progress. Science can be called a journey from the particular to the general, from the local to the universal, from the concrete to the abstract. It's not a steady journey, nor is every thinker suited for it. Geologists are ants exploring the body of a whale. It took us two centuries of observations, after long preliminaries, to learn the Earth's basic structure, piece together its history and test ideas about its evolution.

The preliminaries began in Shakespeare's time, when a scholar named Francis Bacon (1561–1626) argued that "a way must be opened for the human understanding entirely different from any hitherto known." He proposed that unbiased inquiry, based on careful observations and systematic experiments rather than discussions of classical texts, would reveal new knowledge that the ancient Greek and Roman natural philosophers had never grasped by intuition and logic. He also said that one need not be a genius to pursue this new way of knowledge. By recasting natural philosophy as a methodology, he launched modern science.

Bacon called experiments a way of tormenting Nature, of forcing her to give up insights we could never win on our own. It was a cruel analogy from a cruel age, but Bacon's clean-slate, brute-force approach gradually led to useful knowledge. It gave us chemistry. Since then we've learned that another, more far-reaching kind of experiment is to tor-ment our own ideas—to wrestle them and see whether and where they break.

Geology began later in the 1600s with a few simple principles and basic observations. The project grew naturally from scholarly curiosity and the political drive for strate-gic metals, first in the empires of Europe and then in their colonial outposts. We needed to learn the different minerals, classify the different kinds of rocks they compose, and map the rocks of whole countries and continents. Along the way we learned the different fossils and arranged them pains-takingly on a timeline along with the rocks they appeared in. As this complex framework of knowledge began to take firm shape, starting around 1800 there finally came ideas of steadily growing scope that experiments could test.

Scientific ideas, like seeds, need luck to sprout and care to thrive. Many of today's accepted ideas were proposed more than once before winning out—finding examples is kind of a hobby among historians of science. Like seeds, ideas can perish by happenstance, published in an obscure journal or simply ahead of their time. And ideas have enemies: academic rivals, major industries, entire national governments. In this realm, cultural factors matter: the per-sonal qualities of idea-makers, the structure of economies, how societies value knowledge. It takes a canny mind, not just a brilliant and lucky one, to run this gantlet and be heard clearly.

The seeds of ideas finally need the accepting soil of a healthy scientific community that supports excellence. The highest peaks live among rugged foothills, the tallest trees in the densest forest. Mountain View Cemetery has its share of these supporting characters.

Down from the high slopes, on the cemetery's north

side, a van-sized boulder of pure red chert lies penned in its own plot. Next to it, in plot 31, is the unmarked grave of James Graham Cooper (1830–1902), a professional collector of bird and plant specimens who once complained that "the pursuit of science as a private business is a losing game." But other gigs took him into the study of fossils, and he made significant contributions to California paleontology. A more famous fossil hunter, John C. Merriam (1869–1945), is interred in the columbarium below. His UC Berkeley expeditions to the asphalt springs of Rancho La Brea, in the Los Angeles basin, and McKittrick, in the San Joaquin Valley, introduced the sabertooth cat and other charismatic ice-age species to the world. Ezra Carr (1819–1894), in plot 4 without a marker, devoted his geological research to problems of practical interest to farmers as a professor of agriculture at UC Berkeley. Earth science addresses humble as well as sublime matters.

More melange blocks, most of them red chert, crop out among the graves a little farther downhill. My favorite outcrop is in old plot 3, due west of the grandiose monument of Henry Cogswell (1820–1900), a Temperance activist whose connection to geology was his mission to give the common people access to pure, fresh water. It's a shambles of reddish-brown radiolarian chert that has come a long way from its origin, far out in a Jurassic sea, to rest companionably under a redwood tree.

Together, the people brought to stay at Mountain View Cemetery have created a park. The geologists among these dead have done one more thing—they've built a body of knowledge and given us their shoulders to stand on and see farther than our ancestors.

To me, Mountain View Cemetery encapsulates Oakland. Its distinctive terrain of Franciscan bedrock stands out in the East Bay. The beauty of its design embodies one of Oakland's founding values, passed down through the generations. And like the city, it embraces all of its memories and quietly accommodates all faiths and communities.

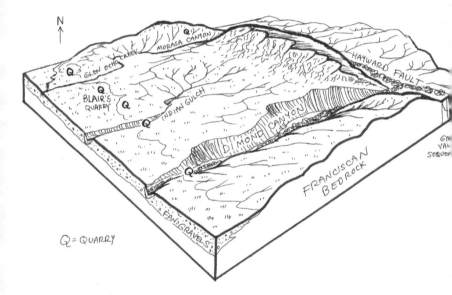

5

THE PIEDMONT BLOCK

Oakland is unusual in having an independent small city, named Piedmont, inside its boundaries. The upscale enclave sits on a set of rocky hills in an unexpected location: west of the Hayward Fault. These hills are a separate block of the Earth's crust, measuring about four by two miles and reaching half the elevation of the high Oakland Hills. The block neatly encloses the city of Piedmont, Mountain View Cemetery and a rim of Oakland territory, also upscale, on all sides. Piedmont's residents voted to incorporate in 1907, two years before Oakland annexed all the land around it. And the little city's fortune, and hence its political insulation, began with wealth and advantages derived from its geology.

No other East Bay city has anything like the Piedmont block. I think of it as a geological mini-Oakland. Both have high rocky rims on their eastern edges that serve as watersheds, both feature springs and streams in their lower lands, and both are related to complexities in the Bay shoreline. Just as Oakland is an assemblage of many different kinds of rock, the Piedmont block contains a variety of rock types. The difference is that in Piedmont, all of them belong to the

Franciscan Complex, the great body of mixed rocks found up and down the Coast Range.

The block has a dramatic eastern border defined by the Hayward Fault, running through Oakland's Montclair district. Whereas Oakland is basically a curving ramp of rock, visible as a whole from any part of town, Montclair is a secluded exception because the Piedmont block, riding north on the fault's west side, hides it from the rest of the city. The one exception is where Shepherd Canyon leaves the Oakland Hills. There a stream crosses the fault and flows straight through the Piedmont block in Dimond Canyon, a rocky gorge with sides that rise four hundred feet. Dimond Canyon is a water gap, a staple of every Geology 101 course.

The landscape in the Piedmont block feels different from the high hills east of the fault with their roller-coaster roads and steep, straight flanks; the slopes here are gentler and more rounded. What's different is the Franciscan bedrock underfoot. It has a unique composition and a distinct story tied to the greater geologic structure of California. And it has influenced Oakland history as much as any mayor, because the Piedmont block generated wealth in hard currency.

High up on the block near its eastern edge, a roadcut on Wood Drive exposes a bit of bedrock. It consists of fine-grained sandstone, a set of hard layers each no more than a few inches thick. But the layers are crushed and warped, as if a bulldozer had shoved a stack of sheetrock against a wall. Although the roadcut is a steep one, no fallen chunks lie at its foot. Despite the disruption, it's integral rock.

The hill is not crumbling, but because the sandstone readily turns into sandy soil, this material is seldom seen. No fossils have been recorded from the Piedmont block, so geologists are vague about its age other than to place it some time in the Cretaceous period, give or take a dozen million years. Around the next curve, though, is a bold outcrop of dark, waxy, wavy-banded chert in chunks the size of trucks. The whole upper part of the block is mapped as melange, the lithic scrapple that mixes widely differing rock types. Franciscan melange, not white Sierran granite or Mount Shasta's black lava flows or even the green serpentine rock of the Mother Lode country, may be California's most distinctive lithology.

The formal description of Franciscan melange, written for the US Geological Survey's map of East Bay rocks, is dense music in a key of many sharps with pan-European lyrics: "Sheared black argillite, graywacke, and minor green tuff, containing blocks and lenses of graywacke and meta-graywacke, chert, shale, metachert, serpentinite, greenstone, amphibolite, tuff, eclogite, quartz schist, greenschist, basalt, marble, conglomerate, and glaucophane schist. Blocks range in size from pebbles to several hundred meters in length." The Piedmont block doesn't have all of these in it, just a dozen or so. And two are here on Wood Drive only steps apart: chert in the outcrop and in the roadcut graywacke, a kind of muddy sandstone whose name, derived from German, rhymes with "Hey Jackie." This is the same terrain seen in Mountain View Cemetery.

Where Oakland meets Piedmont, the road changes its name to Blair Avenue. The border coincides with a topographic saddle and, on the geologic map, the edge of the melange zone. Below, the Piedmont block is mostly a hard, thick-bedded sandstone—rock that was once worth money. Just north of the road, behind old Piedmont Reservoir, is a compact valley with rocky sides the locals call Moraga Canyon. The rock in the canyon and the water in the reservoir represent two aspects of Piedmont's geology it shares with Oakland.

Blair Avenue is a handy thread for this chapter. Its name honors Piedmont's founding figure, Walter Blair, a dairy farmer who arrived from Vermont during the Gold Rush. In 1852 Blair rowed across the Bay to San Antonio Slough and hiked two miles up Pleasant Valley Creek to visit its headwaters in these hills. He bought several hundred acres of prime hillside pasture, built a cabin on a shelf of level ground and began two businesses to capitalize on his land's natural assets. The first was a no-brainer—a dairy farm with grain fields and livestock. The second was a rock quarry that provided essential raw material for the burgeoning city below. The profit from those enterprises bankrolled all that followed: a hotel, a transit line, an amusement park and more.

Piedmont has had at least six rock quarries and Oakland around twenty, far more than one might think. History accounts for this. Oakland was founded near the water in an oak forest growing on ancient sand dunes. The natural landing at the foot of Broadway needed reinforcement with large bare boulders—riprap—to make a sturdy shore. And the young city's loose ground, stripped of trees and cut by hooves and wheels, made for streets that were sticky in the rain and dusty in the sun. The roads weighed heavily on the city's managers. Their decision to pave them by the best approved methods, starting in the late 1860s, called for a basic commodity and lots of it: crushed stone. And the hills of the Piedmont block, an hour away by horse and wagon, had suitable rock and capitalists producing it cheaply as fast as Chinese laborers could dig.

Asphalt and concrete roads were unknown at the time. Concrete was a boutique product made only where bodies of suitable rock, limestone with just the right clay content, could be mined and roasted to make cement. Asphalt, found in natural tar seeps, was used mainly as waterproofing. Cobblestone paving was rare and expensive. Light-duty city roads in the 1860s were built by the macadam method: line a shallow roadbed with a layer of fist-sized rocks, top-dress it with layers of smaller rocks and fine gravel, then compact it all with a heavy roller. Properly made macadam roads were firm and clean in all weather, well drained and easy under hoof and foot. Their main fault was that without regular watering they were dusty in summer. Macadamized roads are gone from today's cities, but they appear in old movies, looking to modern eyes like unimproved dirt.

For the next twenty years, macadamizing the streets kept the quarries of the Piedmont block busy. When Bay Area road-builders switched to bitumen—natural asphalt mined in the oil-rich Santa Cruz Mountains—macadam pavement slowly became obsolete. With further advances in the twentieth century, asphalt and cement could be *manufactured* anywhere instead of mined in lucky places, and that thumbnail history of pavement has played out in every American city of a certain age. Oakland's old macadam rock was probably recycled as landfill in the harbor area and around Lake Merritt.

After Blair's quarry hit its limit and closed, new ones opened in Moraga Canyon and produced crushed stone for all purposes: foundations, railbeds, ground cover, aggregate for concrete. The quarry waste was dumped in Moraga Canyon, and now the upper reach of Glen Echo Creek runs in a culvert beneath the valley's unnaturally flat floor.

The canyon had a happier past as Walter Blair's last enterprise: a garden fantasyland he built in the 1880s named Blair Park. It was a family-friendly place, served by Blair's streetcar line, with a music stage, playgrounds, woodsy paths, scenic gorges and exotic pavilions. A back trail climbed to a height overlooking Mountain View Cemetery called Inspiration Point. The park made extravagant use of the creek to supply waterfalls and cascades. Blair Park was the best thing anyone ever did with Piedmont's natural assets. I think of it as the lost companion to the White Sulphur Springs resort, also in Piedmont one valley to the east—a Hetch Hetchy to its Yosemite.

After Blair's death in 1888, the park's new managers added death-defying balloon rides and aerialist shows, until a few people died and ruined its cachet. Then it fell into the clutches of the Realty Syndicate, which dismembered the property and dug what it callously named the Blair Quarries into the canyon walls. The old main pit, partly filled in, now houses the City of Piedmont's corporation yard.

• • •

The Piedmont block naturally collects water the same way the Oakland Hills do: its elevation and upward slope force the damp Bay breezes to rise as they come off the ocean, cooling until their moisture condenses as fog and rain. That water supply made Blair's dairy farm thrive. The Piedmont block harvests enough water to support five permanent creeks, which come together at Lake Merritt in a compact network shaped like the sketch of a tumbleweed.

The thing to know about streams, from the geologist's viewpoint, is that they're quiet settings except on rare occasions, when exceptional floods do exceptional work. Those events are what shape—and reshape—a stream valley. Maybe once in a century or so, on average, a rainfall event will deliver the maximum possible amount of water. Around here that might be a thunderstorm cloudburst or a "pineapple express" storm, also known as an atmospheric river. Given enough time, every piece of land can expect to experience one. In the geologic present, even the rarest kind of event is inevitable. And a giant rain event turns even the smallest creek into a raging cataract that can excavate its valley in ways hard to picture from daily experience. Every time a storm like that delivers maximum rainfall to the whole Piedmont block, the five streams deliver a quick and coordinated flush that surges through Lake Merritt, especially at low tide.

Geologists with their million-year eyes think of the East Bay as an ice-age landscape with, more often than not, a very low sea level. They see the basin of Lake Merritt not as the slough it happens to be today, where floodwater stops at sea level, but as dry land, the valley of Merritt Creek, where flash floods wash through at full strength. Over repeated ice-age cycles of digging and filling, the catchment of the Piedmont block has helped maintain the unusually deep trough, cut into the coastline and dug straight to the Bay, that now holds Lake Merritt.

Putting aside the extreme events and returning to the everyday, the Piedmont block naturally collects more rainfall than the plains around it. Walter Blair with his farmer's eye surely could tell at once that these hills were a well-watered place.

Water for the first little East Bay towns came from household wells. Larger enterprises needed larger wells, and these were found in the lowlands near the Bay. The best localities became wellfields, owned by small providers of "pure, fresh water," as contracts and news accounts invariably called it. Wells also supplied the first residents in the Piedmont block, and some are still active today, but soon private companies offered more dependable water at low rates. As the East Bay grew, these water companies fought and merged in competitive combat. In this battle, like a real one, the advantage belonged to high ground, where tanks and reservoirs could enlist gravity to supply pressure and regulate the flow independent of the year's weather. Pure, fresh water to fill them was sought in the same high ground.

In 1891, a would-be developer named William J. Dingee, with plans for a district of fine Piedmont homes, applied for a water supply from the Contra Costa Water Company, the provider founded by Anthony Chabot. When they turned him down, he wasted no time. Within a few months, he had the first of twenty-two tunnels bored to tap groundwater from springs in the high, uninhabited Oakland Hills directly to the east in Thornhill Canyon and beyond on the slopes of Round Top.

Dingee was following bad advice. Digging tunnels in high ground is a poor way to produce water. But the water from the tunnels could be piped downhill to a reservoir without using pumps. This limited but cheap supply was enough to get Dingee started. Within two years he was running the Piedmont Springs and Water Company, competing with the firm that spurned him. Piedmont Reservoir, up on Blair Avenue, held the water from Dingee's tunnels. Its high position provided good pressure to the neighborhoods below. Two other reservoirs, of the same type and vintage, sit higher on the ridge in west Montclair, one on Bullard Drive named Dingee and the other, on Estates Drive, named Number One. They were all open basins that got their lids of concrete and steel sixty years later.

From Piedmont Reservoir, Blair Avenue continues west downhill and crosses Highland Avenue, a straight and level route that's rare for Piedmont. Walter Blair's first cabin was here. The town's suburban core is a few blocks to the left. There Blair sold land to a company that built a cultural attraction based on a geological attraction: the swank little Piedmont Springs Hotel near White Sulphur Springs, in the headwaters of Bushy Dell Creek. Blair's horse-drawn stage line brought visitors from downtown Oakland for ten cents.

This enterprise, hotel and springs, was based on the therapeutic use of natural waters, part of mainstream medicine in the nineteenth century. Today we might visit a hot spring or use bath salts for a soak, but taking the waters—drinking or bathing in mineral water for days on end at resorts, on the best advice of doctors and connoisseurs—is a practice long out of fashion in America.

White Sulphur Springs had mineral-laden water, bracingly dank, that attracted fans. The hotel opened in 1871, making the springs an even more popular destination. Mark Twain and other celebrities were visitors. By 1876 a real-estate syndicate had bought more of Blair's land and was laying out the homesites and streets of today. In 1892 the hotel burned to the ground, and without a proper water supply for firefighting, the occupants could only sit in the shade and watch. Water birthed it and lack of water killed it.

What was that mineral water like? Spring waters can be too acidic or alkaline or salty or foul-tasting to drink, but short of that, and barring true toxins like arsenic, their mineral content is pretty harmless.

We aren't attuned to the qualities of water the way people used to be. Sources were all local and had to be assessed with care, even the outflow of pristine springs. The differences in natural waters gave rise to conceptual frameworks as arbitrary as today's belief system of well-being based on mineral crystals. By the mid-1800s, folklore and chemistry had given people a basis to distinguish natural waters by their content of magnesium, calcium, iron, sulfur or sodium compounds. Mineral springs were popular enough to be of official interest to the state government.

The state mineralogist, in 1894, reported that the springs in Bushy Dell Creek "have gained considerable local reputation as medicinal waters." They definitely qualified as mineral water, with concentrations of dissolved solids as great as 1,000 parts per million, ten times the average levels in today's tap water. Iron Spring had a small share of iron and borate, and White Sulphur Spring not only was rich in sulfate compounds and magnesium but also had enough hydrogen sulfide gas to give it a characteristic rotten-egg smell. Both waters had enough dissolved carbon dioxide to tweak the nose and tongue. In 1915, Iron Spring had been re-named Magnesia Spring, and a researcher took flavor notes: "The water of both springs is noticeably sulphureted, and that of the Magnesia Spring also tastes distinctly alkaline."

I have not dared to taste the waters of today's Bushy Dell Creek. I've seen only small and fleeting signs of unusual mineral content in it: rusty stains, bacterial filaments, things like that. Apparently the smelliest sulfur spring was sealed in the 1950s, but natural waters also change on their own. They respond to droughts, floods and the passage of time; their intricate hidden channels can fill with mineral deposits and change their flow. Earthquakes can affect them dra-matically. White Sulphur Springs probably changed often during the seismically active nineteenth century, especially after the "great San Francisco earthquake" of October 1868, at the time the springs first became popular.

Humans have changed the water cycle all over the Piedmont block since Blair built his cabin. We've pumped groundwater, in domestic wells to keep lawns wet and in rock quarries to keep the pits dry; we've turned the vegeta-tion from a carpet of native grasses to a well-watered forest of shrubs and street trees; we've paved over large surfaces, increasing runoff at the expense of recharging the aquifers.

And in still another human-caused change, the climate of California is measurably warmer than in the early days. All of these have tended to dewater the block.

Today the only places that resemble the Piedmont block's original grassland are the manicured grounds of Mountain View Cemetery and the Claremont Club's golf course next door. Within the city of Piedmont, the boundaries adopted in 1907 preserved land of just the right character to attract wealthy residents: not too steep to build on, but high enough to catch the breeze, serve up great views and discourage casual visitors. And so the little city attained its highest and best use—a guarded, landscaped reserve of large detached homes—as the sulfur springs dwindled, the pastures gave way to development, and the stone businesses with their noise and dust left town. The views are still great, though, and the good Bay breezes persist.

• • •

At the lower end of Blair Avenue is the place where Blair opened his rock quarry in the 1850s. Two decades later, under different owners, the pit was abandoned under odd circumstances.

The story goes that in the summer of 1873, a quarry blast broke into an underground spring, which flooded the pit so fast the workers dropped their tools and barely escaped. I don't think so. A "gusher" of that kind is most unlikely, and the newspapers didn't record any such event at the time. To me the account smells of a cover story turned urban legend. Quarries live and die by their costs. As a pit goes deeper, it costs more to keep pumped dry. And at that time, Chinese laborers, and their employers, were being harassed by white vigilantes; California would soon build anti-Chinese policies into its notorious constitution of 1879. It would be a convenient escape for the owners to stop the pumps, let the pit fill up overnight, abandon the tools and the workers, then cover their tracks with a tall tale.

When investigators explored the old pit in 1890, during low water levels in a drought year, the *Oakland Tribune* reported that "even the cars on the tracks are standing as they were twenty years ago, but under forty feet of water." For decades the abandoned quarry was a classic swimming hole, where local children played and where some drowned. Stories swirled about mysterious currents in the water, but wind eddies in the steep-walled pit may have ruffled the surface instead.

In 1996, the city turned the attractive nuisance into Dracena Park, an asset that will endure until the day natural weathering and one last earthquake collapse the walls beyond repair. For now, it's a dramatic place where a sloping lawn rises to a play area at the mouth of a great excavation in the hillside, topped high with trees. Inside, bedrock cliffs half covered with brush guard a round grassy flat. The city regularly adds more fences to keep fallen stones off the walkways and kids off the rocks.

I love quarries, the way geologists do, as scalpel incisions that expose the underworld in detail. I'm grateful for their old, slowly healing scars. The quarry stone from the Piedmont block was well suited for building roads and foundations: hard, durable and consistent. It could be predictably crushed into clean stock and sorted easily into different size grades. One could make steady money with it by keeping costs low.

At Dracena Park the Franciscan sandstone is well exposed, handily fractured into pocketable chunks piling up at the base of the walls. The rock varies from dark, grain-less stone—classified as argillite—to graywacke. In the Bay Area, stone of this type was known in the trade as "blue rock." A slight blue tinge, when present, is typically from pumpellyite, a metamorphic mineral that forms where clay comes under what geologists consider mild temperatures and moderate pressures. Other former quarries exploited the same rock in a discontinuous belt running northwest from here to Novato, in Marin County.

Pulling away a bit, it's fair to wonder why a large block of Franciscan melange and graywacke is sitting here by itself in the middle of Oakland. There are Franciscan rocks around it, but they lie under hundreds of meters of sediment washed down from the rising Oakland Hills. We know this from boreholes drilled in the East Bay and from other tools—gravity sensors, magnetic instruments and seismic-reflection surveys—that yield information about the deep bedrock without our having to bring up samples. The Piedmont block isn't exactly out of place, then; it's been pushed up above its neighbors. This was recognized a century ago, but geologists twisted themselves in knots to explain it, not yet knowing the correct nature of the Hayward Fault.

To my knowledge, no one has published this in a paper, but my idea is that the Hayward Fault forced up the block by pushing it past another immovable object. I'll flesh out that hypothesis in the next chapter.

The same Franciscan stone found in Blair's quarry supported two larger competitors starting in the early 1870s. The Alameda Macadamizing Company dug blue rock a mile southeast of Dracena Park on a ten-acre wedge of land, and a mile the other way the Oakland Paving Company quarried an unusual subvolcanic lava in a larger pit right on Broadway. In 1873 the *Oakland Daily Transcript* bragged that their rock was better than anything in San Francisco, where "on a windy day in summer one cannot more than open his mouth before one's throat is better macadamized than are the streets." For many years in the late 1800s these two outfits were a duopoly that tag-teamed the City of Oakland's road-building contracts. For jobs on the east side of town, the Macadamizing Company would underbid the Paving Company by half a cent per unit of work, and on the west side it was vice versa. The pay was low, but the steady work supported generations of Italian and Irish immigrants who had replaced the Chinese. The frantic years of rebuilding after the 1906 earthquake were boom times, so to speak, for the rock business.

It's hard to imagine loud, filthy quarries operating so near to the parks and estates of the Piedmont block, but in those days dust and dynamite, smoke and noise were approved signs of money-making productivity. Contemporary postcard views of Oakland prettied up the photos, but never erased the smoke-plumed stacks along the Bay shore. The quarries did blasting three times a day, at six, noon and six. As late as 1895, the Reverend Ervin Chapman of the Brooklyn Presbyterian Church could call the detonations "tuneful tones of human progress" and gush that "for years the city of Oakland has been one grand concert hall, in which that music has been daily heard."

But quarries and cities can't coexist for long. Things never got as bad as in San Francisco, where the notorious Gray Brothers quarry chewed up Telegraph Hill for some twenty years, destroying houses and ignoring court orders. Opposition grew to outrage that climaxed in 1914, when a starving worker owed back wages shot and killed quarry owner George Gray—and was acquitted five months later to supporters' cheers. In Piedmont, once the city was incorporated, homebuyers heard a new tune: "The noise of traffic, the clangor of bells and whistles comes faint and mellowed by distance from the city below." The quarries shut down and others opened elsewhere, farther east in Oakland and up in the high hills.

Dead quarries are eyesores with potential. The pits in the Piedmont block have all become civic assets: a city park, a church site, a tennis stadium, a corporation yard, the home of an exclusive school. Each case followed decades of abandonment and involved costly mitigations. A former quarry pit is a challenging space, and the developments that get shoehorned into one have an air of constraint. The high walls of stone impose hard limits on developers as well as certain hazards, such as the risk of rockfalls during earthquakes.

The Oakland Paving Company's former quarry is where Broadway meets Pleasant Valley Boulevard, near the lower edge of the Piedmont block. It first opened in the mid-1860s and for decades it was the largest producer in Alameda County. It had a lot of good rock, and with its dedicated rail spur and convenient location the site was excellent for shipping. The quarry last saw work in 1946, and Oakland old-timers recall childhood escapades among rubble piles in the empty pit.

Today it holds a busy shopping center. Look past those shops, though, and the complex still makes for a jarring scene. On one side, the original campus of the California College of the Arts perches atop slopes of bare rock some thirty feet tall, cut right up to the property line. On the other side sits a dark pool with a sheer rock wall behind it more than twice as high, topped by a fence at the edge of St. Mary's Cemetery. For many years the quarry, the school and the cemetery were active at the same time, somehow staying good neighbors. Some of the quarry workers,

Catholics from Italy, must have been buried near the cliff, where they shook with the daily blasts even in death.

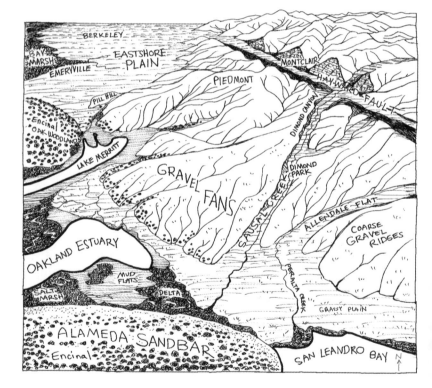

THE FAN, OR THE SECOND LEVEL

The landscape of Oakland has four levels at four different elevations. The buses on the NL line cross the Bay Bridge from San Francisco and enter Oakland on the first level, near sea level. For four miles straight, they glide through the coastal flats of West Oakland, pass the platform of downtown, then sail by the north side of tidal Lake Merritt to its eastern tip. Along the way, the Piedmont block—part of the third level—grows nearer and looms larger against the high hills that make up the fourth level. Then the buses take a hard right and launch into a steady grind up MacArthur Boulevard that tops out at an elevation of nearly two hundred feet.

From here to its end at 73rd Avenue, the NL line runs five more miles across low hills that make up Oakland's second level, touching elevations between about 60 and 250 feet in brief ups and downs. The highs offer peeks of a wide panorama, wooded heights on one side and views across the Bay on the other. The lows are intimate folds dense with family homes, small apartment buildings and little shops.

The hills of the second level cover a long, fat crescent more than seven miles long from North Oakland to East

Oakland, from Pill Hill to Evergreen Cemetery, that bulges out toward the Bay in its midsection. This rounded, sculpted terrain is the city's most charming land, and it all has the same geology. On the geologic map, its outline resembles a tattered Japanese folding paper fan. I call this district the Fan.

The Fan makes for interesting neighborhoods to live in and walk in. The streets curve and slope organically in response to real terrain. The hills are made entirely of ancient firm-packed gravel and sand, bound together with clay. This ragged swath of large sediment piles presents a geological puzzle. My solution to the puzzle of the second level combines all the stories of Oakland's geology I've told so far.

The shape of the Fan on the map looks somewhat like the kind of fan I mentioned before, an alluvial fan. The alluvial fan shown in textbooks is a landform, about the shape of a ginkgo leaf, that a stream creates where it leaves a steep highland, spreads out, slows down and drops the mud, sand and gravel it's carrying. That sediment becomes alluvium—layers laid down in flowing water—as the stream shifts course over the centuries, distributing sediment here and there like a casino dealer tossing out cards.

The larger East Bay streams have built alluvial fans, as creeks do near every mountain range on Earth. But after considering the Fan's geology, I've concluded that its alluvium only happens to form a fanlike shape. Instead, I think the second level was uplifted in a tectonic upheaval during the Pleistocene period. It's part of a larger story with familiar characters: the Piedmont block, the Hayward Fault and Lake Merritt.

These low hills were here long before people came to live among them. They were probably densely wooded during the warm Sangamon interglacial, when high sea levels created the marine terraces at Lake Merritt, but the humans who arrived later turned the hills into open fields. The Ohlones maintained them as a seasonal destination, meadowlands generous with small game, edible bulbs and seed-bearing species. The area had running streams whose valleys, lined with mixed forest and patches of useful shrubs, were parklike avenues between the coastal flats and the high hills. The tribes probably harvested and fired the hilltops early in the year, working downhill toward the streams as the summer drought set in.

The first Spanish exploring expeditions of the 1770s had trouble with the Fan, taking three tries to find the right route through the ungainly terrain. Their records help us see the country as the Ohlones had made it, and the landmarks they noted are still there.

The expeditions came north to explore the territory and make friendly contact with the inhabitants. They took well-used footpaths through settled lands, exploring the way we might "explore" the Napa Valley wine country on Route 29. First was a small reconnaissance mission up the East Bay from Monterey, led by army captain Pedro Fages, late in 1770. On 28 November, Fages sent four men on a sortie from their camp in present-day Hayward to scout the country ahead. When they arrived at the Fan, where the NL line ends today, they probably climbed its southernmost hill, an isolated rise about seventy feet high, to study the scene.

The knoll is now occupied by Evergreen Cemetery, where I can stand and picture what they saw. The view south takes in the wide plain of the South Bay and the high ranges beyond. To the north, though, the land is lumpy with hills, so the scouts turned leftward to walk around their edge. At a spot where these hills pushed close to the Bay shore, they climbed the highest one to reconnoiter, then headed back.

That night they told Fages that they passed "very good and level country," where springs and small streams glinted in the hills to their right.

Fages returned in early spring of 1772 and camped near where the scouts had stopped in 1770, anticipating an easy hike north along the shore. But the next morning the expedition soon found the way blocked by a long slough, today's Lake Merritt. Friar Juan Crespi, the expedition's diarist, wrote that "we were compelled to travel about a league and a half by some ranges of hills, which, although they are all treeless and grass-covered, annoyed us very much with their ascents and descents." Today these hills are known as the Cleveland Heights and Grand Lake districts, and to the residents in their cars the steep ups and downs are merely picturesque.

Once around the slough, Fages reached the Fan's northern edge and looked over the coastal alluvial plain toward the Bay's Pacific outlet. The expedition paused to fix the latitude and record the first observations of the Golden Gate, then moved on. This view can be relived, with some imagination, from the top of 40th Street Way.

Next to visit the Fan was Juan Bautista de Anza's expedition in early spring of 1776. Anza took the way the scouts had shunned in 1770 and proved that the best route up the East Bay led through these hills instead of around them. Behind the lumpy land lay a flatter stretch where the Allendale neighborhood is today, and on the whole the higher route was less hassle. Friar Pedro Font noted that the country was "very green and flower-strewn, with an abundance of lilies."

Today, major Oakland streets follow the old Ohlone roads, the high route Anza took and the low one Fages followed. But the Ohlones themselves, who had greeted the earlier visitors with food, gifts and singing—traditional gestures of outreach to potential partners—were soon cleared off their land by soldiers. The Mission San José priests who ordered that brutal depopulation then set livestock loose on the meadows.

By 1820, when Luís María Peralta gained possession of the central East Bay by royal grant, the look of the land had changed. The Ohlones had been gone for a generation, and cattle roamed freely. The meadows still bloomed but were no longer maintained, and European annual grasses sown by the priests mingled with the native species.

Peralta's family set the first building in their Rancho San Antonio within the Fan, at the western exit of the Allendale plain on a low divide between Sausal and Peralta Creeks. At about a hundred feet elevation, the hacienda was above the mosquito-filled coastal wetlands and away from the marsh air that was thought to carry disease. Visibility of the surroundings was good. The clay-rich soil was right for making adobe bricks. And the best landing on this part of the Bay, just half a league away, was near the hill where the scouts had turned back in 1770. That landing served the Peralta ranch, then grew by the 1840s into the settlement of San Antonio.

As American settlers acquired land in the Fan from the Peraltas, hayfields and orchards and fences began to replace the open range. The hamlets that formed there—San Antonio, Clinton next to it, Lynn up the hill—merged as the town of Brooklyn in 1856. Oakland and Brooklyn, two growing towns separated by San Antonio Slough, were physically distinct: one observer noted that Brooklyn, though lacking Oakland's trees, "is compensated in a measure by the picturesque scenery on every side."

The prominent hill where the scouts ended their sortie in 1770 commanded a view over the Brooklyn landing. On its top the villagers placed a flagpole and a lookout station, to spot approaching ships. The landmark later became known as Independence Square and in 1910 was renamed San Antonio Park. Why is this hill here? Geology offers no answer beyond "it just is."

· · ·

It's tempting, a natural urge, to come up with a story for every small thing. The story I'll offer explains the Fan as a whole but not each feature on it, no more than charting a weather front explains how each cloud fits in the evening sky. To begin, though, as usual we need to see through layers of human history that obscure the evidence.

Early photographs of the Brooklyn hills, taken from Oakland, show the second level as a grassland dotted with a few homes. Although small valleys dimple the Fan's edges in many places, its top surface forms a plateau, sloping gently toward the Bay.

The Fan was outside the city limits for several decades, a hinterland bought and sold in large acreages by farmers and speculators. Then Oakland's capitalists moved in, starting with estate homes on large holdings. When the 1906 earthquake pushed much of San Francisco's population into the East Bay, a wave of streetcar suburbs surged over the Fan, certified free of "Asiatics, Africans and Hindoos." In short order, a carpet of houses replaced the farms and mansion estates that once ruled the second level, although sometimes extra-tall palm trees mark their former sites. The rectilinear street grid faltered in the irregular terrain: in the choicer neighborhoods every street is distinct, and every lot has a special view. Arterial streets trace the former courses of creeks that now run culverted beneath them.

This suburban infrastructure is still largely intact a hundred years later. For much of the twentieth century, a daily tide of white workers took streetcar lines and ferries between homes in Oakland and jobs in San Francisco. The Key System, the regional transit network amalgamated under the Realty Syndicate, used to serve the second level with a good dozen lines, from the 6 College line up Broadway past Pill Hill to the A line out to Evergreen Cemetery. Ghosts of the old rail routes—even their names—can be seen in today's bus lines. Concrete pathways, pedestrian tributaries that once took workers between home and streetcar, survive today as shortcuts for children and other wanderers.

The buses of the NL line follow the Ohlone trail that Anza used in 1776. Old US Route 50, the Lincoln Highway, also took this route in the early 1900s, followed by MacArthur Boulevard in the 1940s and Interstate 580 in the 1960s. Today, whether we drive, ride or walk across the second level, we can still see the underlying landscape and picture how it looked to our predecessors. The eastern, uphill side of the Fan, toward the Hayward Fault, is a string of hills of the third level, most of which are bedrock. The downhill side, toward the Bay, is a variegated landscape of low rises and small gaps through which the Bay sparkles and distant mountains across the water loom, in detail or in silhouette as the weather changes.

This mental exercise is engaging because the swells and slopes and swales—the surface contours of the Fan—are still there. Seeing what's underneath them is harder. What's inside this pile of clay, sand and gravel, and what's it doing here?

The Fan rarely exposes its innards to geologic scrutiny. The streams that cross it are mostly hidden in culverts, and where their channels are visible they're lined with human rubble. Nineteenth-century writers called its soil adobe, in contrast to the loam of the flats. I was not present when Interstate 580 was pushed through in the early 1960s, so I missed the interesting cross sections it must have exposed. Excavations, engineering reports and streamside exposures have shown me that the Fan is made of material of all sizes from gravel to clay. It has obviously been worked by streams, but hasn't been strongly sorted during that process. It's been washed, more than winnowed, downstream. Geologists have mapped the Fan as alluvium, like the material of the flats, but much older. Some of the larger stones in this material have grown soft with age, but that doesn't tell us much.

In the absence of fossils, the Fan's age is, as geologists put it, poorly constrained. Researchers have reported that similar deposits in the East Bay contain freshwater mollusk shells and "extinct late Pleistocene vertebrate fossils," tantalizing words that signify nothing more specific than bones of some sort. Bones of large ice-age animals have been found in Oakland, buried in the flats, but as far as I can tell, none were in the Fan.

On the map, the Fan has the shape of an alluvial fan, but in other respects it makes no sense to call it one. There is no stream at its top, emerging from a mountain canyon and delivering sediment to a proper alluvial fan. Instead a thick blanket of gravel beds spreads up the slope of the Piedmont block, which has been lifted at least a thousand feet above the buried bedrock around it.

This uplifted block and its mantle of Pleistocene gravel are like nothing else in the East Bay. They appear to be a mated pair, which implies a shared history. Boreholes reach Franciscan bedrock in most parts of the Bay, beneath a thick layer of sediment. If the sediment of the Fan was lifted along with the Piedmont block, it should look the same as its deep counterpart elsewhere. Although the borehole evidence is crude, the exposed and buried sediments appear to match.

With that much provisionally established—in fact, it was accepted in the 1950s—I can then ask what uplifted the Piedmont block and the Fan with it. Two facts strike me as clues. One is that the block is not just uplifted but tilted, standing higher on the side toward the hills. The other is that the high side of the Piedmont block butts against the Hayward Fault, where transpression has been pushing up the Oakland Hills for more than a million years.

What happens if we run time backward and undo the movement along the Hayward Fault? The result is that roughly a million years ago, the Piedmont block lay about ten kilometers south opposite the hills of San Leandro. These consist of a particularly robust body of gabbro, a cousin of granite. The San Leandro gabbro is so strong that it deflects the fault by a few hundred meters. And if a similar block of strong stuff on the other side of the fault were forced past the gabbro, the only way out for it would be up. Like a shovel pushed under a buried paving stone, this passing collision of rock bodies would tilt the Piedmont block up out of the ground, lifting the alluvium on top of it as well. I submit, in brief, that the gabbro hip-checked the Piedmont block as it passed. And after that, I think, the block rose some more as the fault carried it past the stubborn body of volcanic rock in Leona Heights, in East Oakland. Since that time, the geologic map shows, splinters of these bodies of rock have been sheared off and strewn along the fault zone.

None of this story is demonstrably false, but that doesn't mean it's true. It means my story is a hypothesis. It's true enough to explain the facts, and it's rooted in natural geologic activities. The tilted block is well-exposed rock at its high end, so I can reason that the gravel that once lay upon it has been washed off during the last million years or so. Lower on the block, the surviving gravel is gradually being eroded, most deeply around Lake Merritt, but the shape of the large mound it once made is clear on the geologic map as a crescent-shaped remnant—the Fan.

A hypothesis, a creative product of imagination, is the engine of science. A good hypothesis, a fruitful one, suggests other questions, or answers to questions, that might seem unrelated. A good hypothesis can find support from other kinds of evidence. Mine accounts for another odd detail uphill in the Piedmont block that I mentioned earlier. The upper edge of the block is a high ridge, except for Dimond Canyon, where Sausal Creek flows straight across the ridge through a notch hundreds of feet deep with solid rock sides. Dimond Canyon is a water gap, a place where a stream seems to have pierced a wall of stone.

Water gaps make sense when we think geologically. The main way water gaps are made, especially in tectonically active California, is that a stream is doing its thing and then the land starts rising under the middle of it, typically on a fault with vertical motion. As the ground is being lifted, the stream bites a gash across it.

Sausal Creek is far too small, too puny a stream to have cut this notch across the rising Piedmont block. Its catchment can't collect enough water to do the work, even during a maximum rainfall event. But consider the Hayward Fault, just above the head of Dimond Canyon. A million years ago, this notch lay farther south and lined up with San Leandro Creek, a much more powerful stream with an impressive canyon in the Oakland Hills. I think this groove was cut as the Piedmont block wrangled with the San Leandro gabbro. And as the fault carried the block away, Dimond Canyon became a tectonically beheaded stream valley—a water gap with an Oakland-style twist.

There are more tests that could add evidence pro or con. Influences from the environment—oxygen, rainfall, cosmic rays, burial and exhumation—leave traces in the rocks. These give us ways to tell when parts of the Piedmont block were first uplifted and then uncovered. Cores can be drilled to map the sediment deposits and trace their origins. Not having the expertise or the means to apply these tests to the Fan, I must leave them as degree-worthy exercises for future graduate students.

· · ·

The Fan also contains a pair of notable middle-scale features. For these I feign no hypotheses, but I can suggest a plausible story.

In two places, avenues of low land, lined with ordinary modern alluvium, have been eroded all the way through the ancient gravel hills. One is the floodplain of Sausal Creek, below the water gap of Dimond Canyon. It's a straight ribbon of low ground, some two miles long and no more than a quarter-mile wide, that rises at a grade of about sixty feet per mile. The floodplain was first colonized by orchardists who prized its flat floor and ready water supply, and through all its changes ever since, the area has been called Fruitvale. What's odd to me is how straight it is—confined on both sides by the hills of the Fan, unable to meander the way that streams prefer. It's easy to picture this linear valley sculpted by floods rushing straight out the mouth of rockbound Dimond Canyon.

The other place, immediately to the east, is more enigmatic. The flatland of Allendale, an oblong patch of nearly four hundred acres, lies at the upper edge of the Fan. Its lower end is partly blocked by a gravel hill, and small creeks flow on its two long sides. Neither of these, Peralta Creek on the west and Courtland Creek on the east, has enough flow to excavate valleys like Sausal Creek's; they barely reach the Bay.

To complicate the picture, mapped near the course of Peralta Creek is a tongue of very coarse gravel, unique in the East Bay, that is said to contain rocks larger than a foot across. Nowhere else in the Fan have I seen a stone bigger than my fist. It appears that an extraordinary event created this deposit by delivering raw material directly from the high hills.

The event I have in mind—an outburst flood—is extraordinary only in the local context. Outburst floods happen in mountain ranges all over the world. The scenario for the East Bay is this: some time in the mid to late Pleistocene, an earthquake triggers a large landslide that blocks a stream. A temporary lake gathers behind this natural dam, the water rises, and months or years later when the lake breaches the dam the whole pile washes away, leaving a ribbon of rubble across the coastal plain. Any evidence that may have remained upstream lies across the Hayward Fault and is now somewhere else.

Such an event could have occurred several times. One could occur today. But to create that tongue of boulders in the Allendale flat, it only needs to have happened once. I have a suspicion, not really a theory, that outburst floods could have helped carve the straight floodplain of Sausal Creek. They're possible anywhere in the fault-ridden Coast Range.

A last notable feature of the Fan is its lower edge. The Ohlones used a level trail along the bayward edge of the Fan, the same one the two Fages expeditions took, and Foothill Boulevard follows that route today. Bus riders who cross the

Fan on the NL line can ride Foothill back downtown on the 40 line. This lower edge, where the second and first levels meet, is a sharp boundary between flat and sloping ground, not the subdued transition one might expect from long erosion. It reminds me of the crisply defined landforms of the Mojave Desert and other arid parts of California, and it reminds me that Oakland too is an arid place in the geologic present. I think the alluvial flats may be burying the Fan as they accumulate sediment, just as the dry lake beds and basins are doing in the desert.

Foothill Boulevard became a major thoroughfare around 1910, when developers induced the city to macadamize the road all the way south toward Hayward. The Southern Pacific Railroad built a new spur parallel to it, one block west. Together, road and rail drove development deep into East Oakland, and the natural boundary became a border where two different development styles met: industries and inexpensive housing on the flats and streetcar suburbs on the Fan.

This wave of population, as it occupied every possible lot in the Fan, uncovered a natural weakness in the terrain wherever streams cut into the Fan and left it with steep slopes. Landslides didn't inconvenience the Ohlones, but the disturbances of American civilization—roadcuts, cellar holes, irrigation, storm runoff and the weight of buildings— made landslides much more likely during the rainy season.

Landslides dot the Fan with scars and empty lots. The largest one, a slump on the north slope of the Sausal Creek floodplain, has been active since 1909. It began as the first homes were built there, and each time it reawoke in a wet winter, destroying a few more houses, was a fresh crisis. After decades of studies and delays, the city bought out the ruined landowners, graded the area and named it William

D. Wood Park in 1976—but the slide keeps breaking new ground at its south end.

The city has always had two minds about Oakland's streams, loving them as ideas but never letting them have their way. Indian Gulch, a stream valley in the central Fan, is a case study in this contradiction.

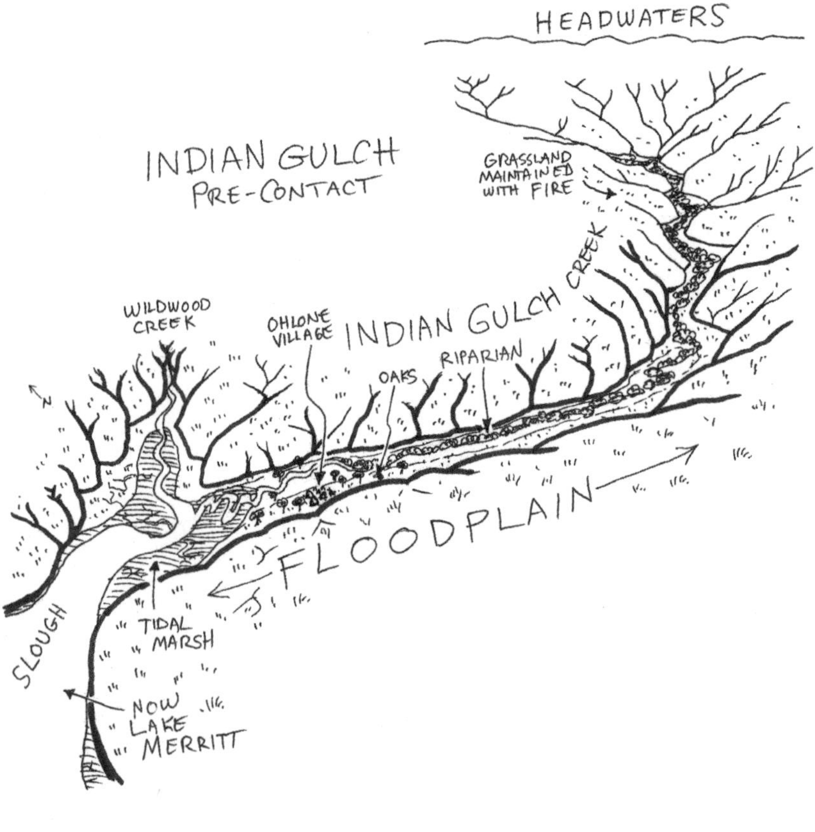

HEADWATERS

INDIAN GULCH
PRE-CONTACT

GRASSLAND
MAINTAINED
WITH FIRE

WILDWOOD
CREEK

OHLONE
VILLAGE

INDIAN GULCH CREEK

OAKS   RIPARIAN

FLOODPLAIN

SLOUGH

TIDAL
MARSH

NOW
LAKE
MERRITT

INDIAN GULCH

Something seems organic about
the treelike, or dendritic, pattern that a stream and its tributaries make on a map. Just as each tree embodies a beautiful
ideal that we call its species, dendritic stream patterns have
a beauty that intuition recognizes but science finds hard to
grasp. Although geologists have long pondered the patterns
of stream networks and stream valleys, the mathematics
to fully understand them has been elusive. But scientists
haven't given up.

The stream valley called Indian Gulch, by Lake Merritt near the west end of the Fan, is Oakland's only natural
feature named for the Ohlone tribes—first by the Mexicans,
who called it Cañada de las Inditas, and again in English by
the Americans. Its current name as a neighborhood, Trestle Glen, was coined by real-estate developers. I single out
Indian Gulch from the Fan's deeper story for two reasons:
it's the most ideal stream valley in Oakland, and its human
history is a case study in the ways cities deal with their natural lands.

Indian Gulch may well have been the Ohlones' favorite neighborhood, and maybe we should be calling it Ohlone

Vale. It had, and lost, chances to become a great city park. Although Indian Gulch has seen many changes and the stream in it has largely been culverted, geologically minded visitors can perceive despite all a stream valley with a soul.

One can sense the valley's form in a hike up Trestle Glen Road from the coastal plain through the second-level hills to the ridge atop the Piedmont crustal block. On a topographic map, the valley of Indian Gulch Creek has the shape of an ostrich plume, an elegant curving stem with short side strands that extends from headwaters in the Piedmont block to the head of Lake Merritt. When I call Indian Gulch our most ideal stream valley, I mean that it has the pleasing treelike form seen nowhere else in town.

The headwaters of Indian Gulch Creek are a rocky bowl where small springs gather in a birdsfoot of gullies and feed a year-round flow. About a mile downhill, two more branches join the creek from their own rocky bowls on its east side. As it descends, more springs and gullies add small tributes of water. The valley grows wider as the creek gradually grows less steep. At its mouth, where Wildwood Creek joins it, Indian Gulch Creek loses its identity in what used to be a marshy estuary as its one-way flow fades to a lazy tidal wash.

A curve that goes from steep to gentle, one that levels off, is called a listric curve. The lengthwise profiles of stream courses describe listric curves; so do certain kinds of geologic faults. Streams form listric curves because they grow their valleys by eroding into landscapes both downward and headward, carrying sediment away as bedrock slowly decays into mud and gravel. Eventually the stream cuts downward as far as it can, to what's called its base level. At that point, the stream reaches a state of equilibrium at that location, with sediment coming in and going out at the same rate, neither building up nor eroding down. Hydrologists say that a stream in this equilibrium is "at grade."

Indian Gulch Creek is at grade where it leaves the bedrock of the Piedmont block and enters the gravel hills of the Fan. There the stream has formed a floodplain, a naturally parklike setting where it meanders in wide curves. There the tribes of Ohlone Vale spent most of their time, and today it's where the dwellers of Trestle Glen spend most of their nights and weekends.

It takes a geological length of time for a stream network to achieve the integrated look of Indian Gulch Creek, and Oakland is generally not a place that sits and waits. Earlier I described how the Hayward Fault distorts the creeks in the Oakland Hills. Over millennia, the fault rips apart and reassembles these creeks, switching the headwaters of one onto the lower reach of another, in a sequence of Frankenstein streams. This also warps their listric curves as their downward flows are delayed by shutter ridges along the fault.

Indian Gulch Creek avoids this tectonic disruption because it lies entirely west of the fault. The creek apparently has nibbled into the hillside ever since the Piedmont block was raised and tilted, about a million years ago. Its watershed of about a square mile was large and stable enough for the creek to carve itself a treelike stream network rather than the cruder patterns, more like twigs, of the smaller creeks around it.

Another thing about streams is that they start in their headwaters as steep freshets of clear water, in beds of bare rock or boulders. Stones and water coevolve as one goes downstream, the boulders breaking and wearing into cobbles that disintegrate to gravel while the water grows cloudy, then turbid with fine-grained silt and clay. When I call

Indian Gulch Creek our most ideal stream, I also mean that it goes all the way from boulders to clay.

A walk up Indian Gulch starts from the eastern tip of Lake Merritt, which was once a wide marsh where clay from the floodplain joined marine clay from the slough. A map surveyed in 1853 indicates that people operated kilns near here to fire this clay into bricks; the wood supply in the stream valley likely made that project pay.

The valley's entrance is an inconspicuous right turn off the commercial stretch of Lakeshore Avenue, an abrupt shift from concrete and clamor to quiet and green front lawns. Trestle Glen Road leads under an elaborate steel arch and into the second-level hills, rising almost imperceptibly over the span of a mile to an elevation of a hundred feet. Along the way the street turns gently leftward, following the organic curve of the valley and its stream. The creek curves more widely than the road, appearing in backyard glimpses first on the right, now on the left.

Once within the valley it's clear why Spanish speakers called this place a *cañada*, or glen: an open wooded ravine with a flat floor and running water. Old photographs show a grassy floodplain along a permanent stream, dotted with large trees. The valley's walls held back the Bay wind and Pacific fog, and its forest offered summer shade and winter shelter. The oaks in the glen sustained the Ohlones with acorns. The marsh and the shore were in easy reach. The people bathed in the creek after sessions in the sweathouse. And no less important, the glen offered amphitheater-like settings for ceremonial activities to mark the year's cosmic cycle and important occasions of life.

The glen, named for its prominent village, kept the Ohlones very well. But the European occupiers were blind to its legacy; to them the oaks were merely wood and the acorns fodder for swine. The Ohlone legacy was not a building or an institution but the land the explorers encountered: flowered plains and ridges, lush wetlands, a shore teeming with plant resources and game, sea air laced with cleansing smoke. The tribes were not ghosts in the landscape, squatters in a wild Eden ripe for taking, but caretakers of a realm that sustained them. They cared for this place in every sense.

What followed the European land seizure was the American process of residential development. We take it for granted, but I sometimes call it landslaughter: a living, inhabited landscape killed like a steer and its life blood drained; its hide and tallow rendered into money; its flesh surveyed and cut into parcels, marketed in tranches and laid out for sale on the courthouse steps. Here the result was to remake Indian Gulch, the Ohlones' favorite neighborhood, into Trestle Glen, one of Oakland's favorite neighborhoods, while covering the tracks. Nevertheless, the geologist is trained to see past these labels and perceive the setting beneath.

Many people find that something at their doorstep, in the natural world, catches their attention and sparks a passion. They come upon a trail and follow it. For some the journey starts with the birds, with others it might be the native plants, and with me it's the rocks and landforms. As our quests take us deeper, we find ourselves beginning to see the civilization behind the city—its demands, its costs and its consequences—from outside. This valley, during its passage through human history from village to district, has kindled delight, disgust and the land-lust of developers. Indian Gulch displays the losses and gains of landslaughter well. The least I can do, as I tell its geological story, is keep its old name alive.

• • •

The primal crime of the European conquerors was their Doctrine of Discovery—ownership by simple decree. "I claim this land in the name of the king," I would recite as a child at play in the fields, and like much child's play this little speech has a violent core. The New World was seized from its aboriginal people in the names of European royal families, oligarchs who ruled by force of arms and authority of the cross. Juan Cabrillo uttered a similar formula to claim California in 1542 for Charles I, the Spanish king of the Habsburg family. After the Ohlones were abducted to the missions, the king of Spain awarded much of the East Bay in 1820 to a retired soldier, Luís María Peralta, for services rendered in that human roundup. Peralta gained the title of *Don* and a homeland for his own family dynasty that he named the Rancho San Antonio.

The four Peralta sons moved onto the East Bay grant, started livestock ranches on the wide coastal plain, and raised families. When the mission system ended in the 1830s, many of the tribespeople found refuge in servitude with the Peraltas. Some of the Ohlone servants and herders must have recognized their former home, unpeopled for a generation; perhaps their stories led to the name Cañada de las Inditas. When the Americans dismembered the Mexican land grants, the Ohlones were cast loose, but never exiled. Even without their land, they held their culture close and began a long struggle for recognition.

When Don Peralta divided the ranch among the sons in 1842, he drew a boundary up the centerline of this valley to separate the shares of the two middle sons, Vicente's on the north and Antonio's on the south. In 1852, the county government separated Oakland and Brooklyn Townships here, using its new English name of Indian Gulch. These foundational acts of ownership split the valley down its axis, and few land parcels ever crossed the creek.

As Oakland grew, residential developments crept over the second-level hills to the valley's rim, but the floodplain was left alone. It was valuable as good pasture on rich, well-watered soil; it was also hazardous during rainy season when the creek poured mud over its banks—the geologic process that makes floodplains. Developers favored hillside lots for their drainage, the views they offered and the breezes they enjoyed.

Landslaughter, among other things, pits the owner's profit against the public's benefit. As residential districts expanded over Oakland starting in the 1890s, civic visionaries began urging the government to set aside attractive lands for public parks. Indian Gulch was the prize in an early bout between public and private good, and after several bruising rounds the subdividers won it. Things could have gone differently.

Its distance from downtown, its divided ownership, the difficulties of sewage in the floodplain, business fluctuations and a dithering city government left Indian Gulch a mostly unbuilt place, a rural valley surrounded by hillside subdivisions, until around 1920. It then became the very model of a modern Oakland suburb tract, complete with the racist exclusions of the time. The houses in the glen have a similar size and style, due partly to the era they were built in and partly to building restrictions written in the conditions of sale. With utilities undergrounded, the streetscape's only decoration is glass lamps on ornate steel poles. The homes are closely framed by native tree species on the slopes behind them. The creek is buried out of sight except in a few backyards.

The neighborhood along Trestle Glen Road is cozy and scenic, like the set on a stage. The parade of period homes is both impressive and soothing. Looked at another way, the neighborhood is a real-estate monoculture, the kind that's tightly planned and quickly laid out as geometric lots in former pastures and other greenfield settings—suburban sprawl, occupied space, created for fast profit.

When packs of American predators brought down the Peralta family ranches, Indian Gulch was a prime cut of the carcasses, and not just because it produced good hay. As the city grew into its outlying lands, a dozen or so large landowners placed bets on the Oakland game board, eyeing one another's moves and sniffing the economic winds for advantage. Their typical gambit was to push a streetcar line into one of their holdings, subdivide it into tempting lots and market its natural advantages—the views to be had and the availability of fresh air and good drainage. An 1869 map shows the glen's northern side laid out as "Lake Park," but that was just a developer's vision that involved a streetcar line from the north, through Piedmont; the development never opened, the streetcar line never materialized, and no trace of it is left today. Lake Park was too far out of town.

In the 1870s one of Oakland's wealthiest citizens, the private banker Peder Sather, acquired the Lake Park parcel, and for many years he leased it to a farmer. After Sather died in 1886, his executors bided their time with the property. The Sather tract and the two adjoining parcels upstream, owned by founding father Samuel Merritt and railroad baron Charles Crocker, sat undeveloped.

As one follows Trestle Glen Road upstream, the valley floor remains flat, but it gently narrows and widens. The street's even leftward curve occasionally brushes against a steep slope. Gradually the floodplain turns narrower, but approximately at Grosvenor Place it opens fairly wide, a fit land for a thousand dancers.

This was the original Trestle Glen. It got this name on 23 April 1893, when Trestle Glen Park opened on the south side of the creek, its wooden pavilion built to hold up to a thousand people for concerts, dancing and camp meetings. The park was owned by the electric East Oakland Street Railroad, which ran a spur north up 4th Avenue to Park Boulevard that mounted a low ridge and then sailed across Indian Gulch on a long wooden trestle, forty feet high. Like gondolas in today's amusement parks, fancy double-decker railcars glided at treetop level, their occupants seeing and being seen.

The plan was to push the rail line through the Sather tract and on to Piedmont later that year to recoup the railroad's large investment in the trestle. But the line never went farther. The Panic of 1893 upended everything. It lasted four years and was the deepest depression America had ever suffered. During that time, an outfit called the Realty Syndicate bought a majority share in the railroad and scooped up the whole operation. The Realty Syndicate was the largest land dealer and developer in Oakland's history, and the Trestle Glen deal was one of its first.

During these years, talk turned to the need for new parks in Oakland. What existed were horse-racing tracks and private sporting grounds, along with proto-Disneylands like Blair Park in the Piedmont hills and Shellmound Park at the mouth of Temescal Creek. What was needed, said thought leaders of the City Beautiful movement and other advocates, were naturalistic spaces with mild elevations and inspiring views, where all classes of citizen could recreate in moderation. The ongoing depression brought opportunities for Oakland to acquire suitable lands. The city could raise the money by having its voters approve a bond issue.

In 1896 the city's Board of Trade issued an ambitious plan proposing a wide belt of parklands from Lake Merritt to the high hills, encompassing about two square miles in total. "The most central and available location for a public park is undoubtedly at the head of Lake Merritt, known as the Sather, Merritt and Crocker tracts," it stated. "This would include Indian Gulch, Trestle Glen, all of the Sather and Merritt hill and valley lands and a small portion of the Crocker lands, making a gem of a natural park of about 400 acres that would attract all citizens and visitors."

What was considered desirable in such a park? Reports of the Board of Trade plan were vague about the attractions. Capitalist James Gamble offered a bare list of the property's virtues in the *Oakland Tribune*: "It has sheltered nooks, a wooded glen, a running brook, rolling hills, which reach to the Thermal belt, where orange and lemons grow to perfection, and from its highest point a magnificent view can be had of the bay of San Francisco and its surroundings." Perhaps he pictured grounds resembling a manor estate in this "natural park." What average citizens wanted in a park, though, was a pleasure ground, like Lakeside Park and the regional parks of the century to come. That would mean improvements, and those would cost the public money.

The park proposal started to look real a year later in 1897, when the Sather executors offered Oakland the Indian Gulch tract at a take-or-leave price they thought fair. The Realty Syndicate pledged to add the Trestle Glen parcel as a gift. Within days, however, a half-dozen other landowners offered their own properties as bargain parklands. During the political whirlwind that followed, the *San Francisco Examiner* quoted a former city attorney's complaint: "We want a park, not a big ranch which we never can reasonably improve. The Sather tract is nothing but a big ranch." Voters, made wary by the long depression, seemed to have little appetite for a bond issue. Negotiations fell apart by Christmas.

The following spring, a fire destroyed the pavilion at Trestle Glen Park. Barely two weeks later the streetcar line halted service; the Realty Syndicate complained it was losing eight dollars a day. In truth, the syndicate had acquired the competing private parks and felt no pressure to rebuild this one. The San Francisco *Call* wrote, "Now the whole enterprise will be abandoned as the syndicate, having no opposition, can divert all the summer pleasuring to Leona Heights or Blair Park." The streetcar line resumed service and events continued at Trestle Glen, but talk of a larger park in Indian Gulch ceased.

From my viewpoint, that was just as well. The idea of the park sounded good. The three parcels enclosed most of Indian Gulch on the inner side of its graceful curve—but that was only one side of the ostrich plume. With the creek split lengthwise, it would have been a strange park by today's standards. Don Peralta's dividing line was to blame. Had more land across the creek been part of the deal, the park might have preserved the watershed and biome of this ideal stream valley largely intact. Instead the owners kept their holdings.

· · ·

In 1904 a new round of proffers and struggle took place over the same would-be park. Again the *Examiner* railed against the plan, accusing the agents in the deal of self-enriching "jobbery" and calling the valley "a rugged canyon, of revolting appearance, and totally unfit for such a purpose."

Perhaps the editors meant the upper part of Indian

Gulch. Above the Trestle Glen floodplain, the valley narrows and Trestle Glen Road has fewer side streets lined with homes. Along the way the valley becomes a true gulch: a narrow, steep-sided ravine. Imperceptibly the way grows steeper. Abruptly at Creed Road the country changes, and shoulders of bedrock emerge on the valley's flanks. The homes loom taller, rearing up against and leaning into the slopes. The street itself narrows. Bedrock crops out by the sidewalk—blond Franciscan sandstone of the Piedmont block. The remains of a small quarry lie by a paved pathway to Elbert Street that Trestle Glen residents once used to catch Key System streetcars. Somewhere around here too was "a very retired and secluded locality" where a factory sat in the 1880s, extracting the oil from eucalyptus leaves for medical uses and as a rust preventative in boilers.

Calling this place a rugged, revolting canyon was an overstatement, but one well within the range of newspapers in the age of William Randolph Hearst. And indeed Trestle Glen Road soon turns up a side ravine to Park Boulevard, as if giving up on Indian Gulch. Probably the Ohlones took this route out too, leaving the upper part of the creek to itself.

Farther upstream in the little city of Piedmont, the living creek bustles in a rocky gorge, shared by landowners at the rear of their lots. During the wet season, its tumbling rapids can be glimpsed in places and its waters easily heard from the streets above. The two main branches of the stream join on a shady three-acre lot, the largest in Piedmont, that includes more than two hundred yards of creek bed. Here their combined flow has cut the canyon especially deep.

The nearest thing to the original valley still left is the grassland patches around the towers of an electric power line running up the gulch's south flank. Because the line is a century old, built by the Great Western Power Company when the city had competing providers, these tower pads could conceivably have traces of the Ohlone meadows that once overlooked the scene.

In its upper reach, Indian Gulch Creek consists of three forks. The streets there were laid out to optimize the division of land into streamside lots. The uppermost catchment, high on the Piedmont block, is no longer a place of streams but one of cliffy outcrops and dramatic views. The highest spot on the rim, Pershing Drive, is graced with large outcrops of greenish Franciscan chert, part of the melange belt. Just below, Somerset Road winds along a wall of rock exposing pale Franciscan sandstone.

What if Indian Gulch's park had been all it could have been?

Imagine strolling along the winding brook, lined with oak and bay and black walnut trees. There would be a space for public performances in the Grosvenor Place flat, with interpretive signs about the gilded streetcars and dance floor at old Trestle Glen Park. Tennis courts and a garden center, surely. Perhaps a fish hatchery. Today the Ohlone legacy would be acknowledged with a set of signage, at the very least, and maybe a replica village like the one in Coyote Hills Regional Park farther south near Fremont.

Pathways up the gulch would lead to the rocky headwaters, and benches for contemplating the view. Rock climbers would patronize the rugged heights. Trails at the park's upper end would connect to Dimond Canyon and Montclair. At the same time, one or two roads might need to cross the valley. Picture them vaulting the creek on elevated bridges, homages to the old trestle.

The land would have to be managed, though not in the Ohlone way. Fire would be ruled out. Dams and lakes would be put in place for flood control. Deer would make homes there; coyotes and mountain lions would visit too. Poison oak would be discouraged in favor of charismatic wildflowers. There would be lawns to mow, brush to chip, trees to prune. There would be crime to address and public wear and tear; also cultural events and groups of benevolent volunteers to supervise. It would be a city's natural park, with all the creative ferment and compromise those words imply.

If I were in charge of the geology-related interpretive signage for Ohlone Vale Park, I'd use it to talk about the watershed, the floodplain, the quarry and the sandstone cliffs. I'd also point out the remarkable set of slickensides— polished rock surfaces made by faulting—exposed near the east end of St. James Drive, and I'd even lead walks.

But stop the daydream. Oakland voters probably recognized that it takes a lot of work to civilize wild land— to improve a big ranch. For instance, Central Park in New York City took decades of serious geotechnical engineering, and the replacement of most of its topsoil, to realize the Olmsted plan and make it resemble a mid-Atlantic pastoral paradise. Golden Gate Park in San Francisco, likewise, took many years of effort, starting in 1870 under superintendent John McLaren, to turn a thousand acres of windy dunes into a gem of a natural park.

Oakland had more chances to acquire portions of Indian Gulch in 1914, 1917 and 1919. But after the upper gulch became part of the new city of Piedmont in 1907, the park vision was never the same. Over the years the valley became the secluded scene of waylayings and suicides; in 1909 the *Examiner* called it "a spot where several human tragedies have taken place." The size of the park proposals gradually dwindled to ninety acres. Another developer bought the Sather tract in 1913, and the process of landslaughter resumed in the glen to its current endpoint: a long set of detached, single-family period dwellings with green and tidy grounds. In 1915 the influential city planner Werner Hegemann, in his elaborate development plan for Oakland and Berkeley, called the failure to acquire Indian Gulch an "unredeemable mistake"—yet the resulting suburb was precisely the kind his plan idealized.

Real-estate economics has a concept that governs property appraisals called "highest and best use": the combination of feasibility, legality and maximum value. The weasel word in the formula, *value*, is defined purely in terms of money, but land seeking its own level of value isn't a natural law like water seeking its own level or streams a listric curve. Value is a matter of policy, guided by custom and the culture that supports it. Landslaughter can be changed.

The upper valley, in the gulch, reached a different endpoint. Instead of rows of kindred homes, Piedmont favored large specimen houses and wide-lawned mansions for which the only word is baronial. There are stairways and footpaths like those in the streetcar suburbs below, only more substantial and ornamental in keeping with the wealthier milieu.

Piedmont succeeded in placing a public park, though not a natural park, in Indian Gulch. Hampton Park, dedicated in 1974, is a couple acres of perfectly level landfill devoted to various sports. The middle fork of Indian Gulch Creek runs beneath it in a culvert. A small pond above the park, a former reservoir named Tyson Lake, is a private backyard amenity for a circle of about a dozen houses. The entire east fork is now underground. And the main fork, on the west, where the old Mexican ranch boundary is still marked on the map, is an anonymous trickle accessed only by a few dead-end streets.

De-natured throughout its length, Indian Gulch has

been transformed from a sustaining landscape to a pictur-esque stage set for American suburban life. But the shape and geology of this much-changed valley are still there to be seen behind the scrim.

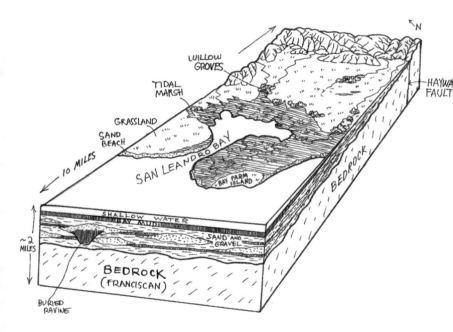

8

THE BAY SHORE AND FLATS

The Bay shore and the flats may be easy to overlook these days; they can be hard even to see. What interests the geologist is almost fully hidden. In their ways the Oakland shoreline and the adjoining coastal plain contributed the most, and gave up the most, in support of Oakland's development. That human history has subdued them, but not completely and not forever. Geology abides.

Oakland wouldn't exist today without the good soil, good water and good building ground in the flats, or without the abundance of sand and wide tidal lands on the shore. I'm treating the shore and the flats as a single subject because together they constitute Oakland's first level. Below the ground, too, the flats and the shore are one domain. On the surface, they're quite different and deserve their own introductions.

The flats extend past the city boundaries all the way up and down the East Bay. They're basically a blanket of sediment hundreds of feet thick, washed down from the hills to, and into, the Bay. Every stream contributes to that blanket: in the speeded-up movies geologists have in their heads, a thousand years a second, the streams are loose fire hoses that lay down the blanket by shooting mud in all directions.

This mantle of sediment rests on Franciscan bedrock, the same kind that crops out in the Piedmont block. The land has an easy slope, rising roughly a hundred feet with every mile toward the hills.

The natural shore of the East Bay is mostly a straight stretch of marshy lowlands and shallows between two rocky bookends, on the north the Potrero Hills west of Richmond and on the south the Coyote Hills west of Fremont. At its midpoint, the shore is interrupted by three ice-age dune-fields, made of Merritt Sand, that bulge into the Bay. From north to south, a fat one underlies downtown Oakland, a slender one hosts the island city of Alameda, and a little one is the nucleus of the Bay Farm Island peninsula. Between them, these three dunefields enclose two arms of the Bay: between Oakland and Alameda is San Antonio Creek, called the Estuary today. Between Alameda and Bay Farm Island is San Leandro Bay.

The line of the shore, between water and land, is a profound boundary, but also a superficial line that shifts dramatically over geologic time. What's important is the buried record left by those shifts as the sediment has piled up over the last million years or so. The underground structure of the shore and flats contains natural wealth in the form of groundwater, which underwrote Oakland's growth into a major city. At the same time, Oakland has overwritten almost everything the shoreline and flats used to be. That human history is where I'll start.

The shore of San Francisco Bay used to be wide wet-lands. Friar Pedro Font, taking notes during the 1776 Anza expedition, was not impressed: "The greater part of the shore of the port, as I saw it when we made the circuit of it, is not clear, but miry, marshy, and full of ditches, and is consequently bad." But to the Ohlones, that shore anchored their lifeway: the mire was mudflats rich in shellfish, the marsh a storehouse of foods and materials, the ditches a network of water-paths for their reed boats. The Spanish colonists and the Mexican ranchers mostly let the coastal wetlands be, except to build small landings for their vessels. The Americans at first took rowboats into the marshes, hunting ducks to sell for a dollar apiece in Gold Rush San Francisco.

Then the city-builders of Oakland set to work and reconstructed the shore into something Friar Font would have praised. They drained the mires with grids of ditches, built solid land upon the marshes and remade the soft shoreline into deep-water channels lined with stone and concrete—a herculean, century-long program with the bland name of land reclamation. On Oakland's reclaimed "made land" the builders put a major city's vital organs: railroad yards, a harbor, freeways and an international airport. They put military bases and whole new residential suburbs on made land. They turned a fractal habitat to a maze of linear seawalls. Although many American cities have used reclamation to build themselves new room, few have gone so far.

The destruction of its quiet primeval shoreline was the largest environmental injury Oakland ever inflicted in its dogged climb to civic greatness. To the Americans, the marsh was something to be made into deep navigable water or dry developable land, one or the other. Dredging the Bay and burying the marsh with the spoils, they created made land and made water in one act that forced a binary scheme upon an organic, gradational landscape. This work ended only when the marsh was almost gone. Few photos of its original state survive.

The broad coastal marsh was an old-growth shoreline, stable for thousands of years, that wove plants of all kinds and animals both migratory and sessile into a living fabric without large, iconic species. Offshore were tidal mudflats full of clams and mussels. The coastal tribes built large mounds of their discarded shells all around the Bay; the shellmounds also served as campsites and burial grounds. Early Oakland maps show an Ohlone shellmound by the mudflat at the head of San Antonio Creek, where Livingston Street meets Embarcadero today, and at least two others once stood in Alameda.

The Americans used the mudflats to keep racks of native oysters brought down from Washington, harvesting them on demand for Gold Rush oyster bars. Later the railroad delivered barrels of Atlantic seed oysters for maturing in the Bay. Pollution from sewage ruined the local oyster farms around 1900, but not before a young Oakland writer named Jack London memorialized them. The days of utilizing the living Bay ended.

San Antonio Creek was named for the Peralta ranch. Don Antonio Peralta built piers at a landing there by the foot of today's 14th Avenue. The geological setting was the same ice-age marine terrace, about six meters above sea level, that I described at Lake Merritt. In the 1830s the Mexican ranches began trading hides and tallow for manufactured goods from British and American merchant ships anchored off the little town of Yerba Buena, not yet called San Francisco.

Although vessels came and went at Peralta's landing, San Antonio Creek could be tricky sailing. Most of the natural channel was deeper than ten feet, but where the creek met the Bay was a shallow barrier of sediment—a bar—where the water was less than knee deep at low tide and the bottom was sticky mud. Bars form wherever streams reach the coast, their current slows, and the sediment they carry falls out of suspension. A similar bar, many times larger, lies off the Golden Gate.

Peralta's landing was the main shipping point during the 1840s redwood rush, and the village of San Antonio grew around it. When the three American squatters came in 1850, they fixed the foot of Broadway on the sandy bluff west of the slough. The new landing was their ticket to prosperity and the bar their prime impediment. For their scheme to work, they needed to dredge a deeper channel through the bar. Everything grew from that.

At this point I need to tell about the squatter Horace Carpentier's greatest swindle. In 1852, after he had arranged for the town of Oakland to be incorporated on land he had no title to, the trustees at their first meeting granted Carpentier the town's entire waterfront. In 1854, Carpentier became mayor of the newly incorporated city of Oakland with more votes than the city had voters. Soon he gained the ferry franchise too. In 1868 he transferred the waterfront land to a partnership with the Central Pacific Railroad, which held this monopoly against all comers for forty years while everything related to the waterfront proceeded on the railroad's terms.

• • •

Dredging began in 1859 and never stopped, as Oakland gradually remade its shoreline into a major harbor. There was always more to do. As sail power turned to steam, each generation of larger ships needed bigger channels and sturdier wharves. Starting in 1874, the work followed a long-range plan to build what nature had half made—a grand harbor fit for the commerce from the new transcontinental railroad.

The harbor plan—channels, basins, groins and seawalls

—occupied the Army Corps of Engineers for thirty years. Its culminating stage, completed in 1902, was a mighty channel dug across a mile of dry land, four hundred feet wide and eighteen feet deep at low tide, from the east end of the Estuary to San Leandro Bay. The idea was to create an artificial strait for the tides to sweep the harbor clean of sediment. The Tidal Canal turned Alameda from a peninsular town into the Island City of today.

This work completely changed the Estuary. No longer the living estuary of San Antonio Creek, it's a channel with no shallows, no marshes and no mussels. Its floor is kept clean, and its banks are walls of concrete or slopes of riprap, brought by barge from local stone quarries to armor the sandy shore. I always give the riprap a look. In Middle Harbor Park, a replica pier recycles vintage riprap from the 1870s in a display of mixed Bay Area rock types that I find charming. The newer stuff, from more distant quarries, has better durability but less variety and character.

Although the Tidal Canal was the last project to increase the area of made water, land-making continued for decades. The railroad lost control of the waterfront in 1910, when the court ruled that Carpentier's land extended only to the low-tide line as of 1852 and that the made land beyond it was state property. In 1911 the city annexed all of the state's share, including the submerged land under the Bay out to the county line. Finally, in 1925 the independently governed Port of Oakland began to develop this huge trove of public land.

Made land was manufactured in two ways. The first was to dig ditches and drain the marshes enough to support hayfields and pasture. Around Oakland this drained marshland was remade later by the second method: build a barrier around a parcel, then fill the resulting basin with dredging spoils. The railroads led off by pushing piers and causeways far into the Bay, where ferries would take the cars the last mile to San Francisco. These spikes on the shoreline formed the spines of growing harbor complexes in Oakland and Alameda, all built on made land. The material for the new land typically came from enlarging the ship channels, which today are as deep as fifty feet, but construction waste and municipal garbage were sometimes part of the mix.

Reclamation added dramatically to the harbor lands. Alameda doubled in size on aprons of fill, including Government Island in the Estuary. In Oakland, landfill projects pushed West Oakland's shore a mile into the Bay. The Key Route Basin was the largest, a four-hundred-acre piece of tideland that the Realty Syndicate surrounded with rubble and pumped full of fill in the 1920s. It became home to the Outer Harbor complex. Bay Farm Island, once a small patch of sand in a large marsh, was turned into a peninsula with enough land to open both a spacious airport and a golf course in 1927, a business park in the 1950s and a residential development in the 1960s. The airport, with its steady winds and freedom from fog, attracted barnstormers and rock-star fliers the likes of Amelia Earhart, Lester Maitland, Albert Hegenberger and Jessie Coleman. It made Oakland a complete twentieth-century city while erasing its largest coastal marsh.

Land-making climaxed in the mid-1900s as people floated plans to remake the whole Bay. The most ambitious one proposed to turn San Bruno Mountain into crushed stone and dike off all but a fraction of the Bay around the port of San Francisco. Oakland's example of headlong, wholesale reclamation inspired these people, and dismay at the outcome inspired others to advocate restoration. Encroachment into the Bay stopped in the 1960s and restoration began;

even in Oakland, shoreline marshes are being restored from remnants or built from scratch—made coast.

Although the plants and wildlife seem to approve of the new marshes, made land isn't natural land. To function well as good ground, landfill material must be chosen with care and procured, mixed, compacted and maintained by people with expertise. Nevertheless, one can't replicate in a week, with dump trucks and dozers, a natural tissue laid down grain by grain over centuries. Most of Oakland's new land was built to the standards of the times, but the Hayward Fault hasn't tested it yet. Much of it, or the natural ground beneath it, may soften or liquefy under intense shaking.

Dredging has never stopped, because it must continue as long as sediment drifts into the ship channels. These days the spoils generally go to other parts of the Bay, to build new inlets where former hayfields and salt pans are being re-reclaimed as tidal marshland. Elsewhere offshore, clean Merritt Sand is dug up to reinforce beaches. As the sea continues to rise on a warming Earth, this valuable material will bolster the shore in more and more places.

• • •

The three ice-age dunefields that interrupt the East Bay shore are almost invisible today. The northern one is now Oakland's downtown. Bay Farm Island, once famous for its vegetable gardens, is now a paved suburb of Alameda with no perceptible topography. The middle one, Alameda proper, still displays a few distinctive geological features.

Alameda entered history as a forested peninsula where the Ohlones came to harvest acorns and feast on shellfish. One of their shellmounds, where Mound Street runs today, was the highest point in town until it was leveled in 1905. The Mexican ranchers visited mainly for lumber and firewood from the Alameda encinal. The Americans found the peninsula especially well suited for railroads—its western tip was the solid ground most convenient to San Francisco—and for lush Victorian estates, masterpiece theaters of the Gilded Age suburban lifestyle.

Relentless reclamation doubled Alameda's area of dry land and smothered its marsh, while the Tidal Canal made the peninsula an island. The north side, facing the Oakland harbor, became a shipbuilding district. The western end was built up for military bases and an airport. On the south side, the sandy mudflats became a long tract of 1950s-era homes. There the original shore is marked by a small change in elevation and a string of backyard lagoons, stranded when the land was extended with fill.

The oldest maps show bluffs of sand—Merritt Sand—a good ten feet high along the bayward shore. This Pleistocene dune sand fed a long beach, manicured to a tropical ideal by the gentle Bay surf. It supported a string of resorts where crowds enjoyed the best bathing beach in the Bay Area. The bluffs were there because even the mild Bay waves were steadily cutting into the land. One old resident told a reporter in 1892 that during the previous fifty years "the sea has eaten away at least 200 feet of the land at the foot of Park Street." This would explain why some of the Bay's shellmounds stood offshore: the Bay had carried off the land around them.

The coast is retreating under the sea's attack not just here but throughout California. Everywhere in the world that coastal cliffs exist, they're crumbling into the surf, which rakes and winnows the wreckage and directs most of it offshore. Nevertheless, the sea has not won; something keeps restoring dry land. Geology can explain why Earth abides: the tectonic forces that keep continents high play out tens of miles below us at a scale perceivable only on maps, in slow interactions of great slabs of the Earth's outer layer. Our landscapes—the places we live in—are evanescent skins on deep doings.

• • •

As for the Oakland flats, human activity has left the coastal plain, like the shore and dunefields, a ghost of its former self. To the Ohlones, the flats were a vast meadow laced with small streams and dotted with willow thickets. They used seasonal burns to suppress trees and brush in favor of desirable plants and grazing animals. Their practice was a landscape-sized version of the carnival trick of keeping a top spinning with a whip, applying force in expert flicks to maintain it in energetic metastability. The key to managing the Ohlone ecosystem was their attentive and disciplined culture. When the Spanish occupation turned the flats to open livestock range, the habitat wobbled off its metastable state into a boom-and-crash biome of introduced seasonal grasses.

The Americans, alert for profit above all else, loved the flats for their potential. "O, what a beautiful plain for a city," one writer put it in 1868: "four miles and a half back to the mountains." They easily imposed a series of land uses on the flats—grain farms, then industries, then family homes—each one extracting a new tranche of wealth, first from the soil, next from the ground, finally from geometric lots on grids of residential streets.

Whatever was planted in the East Bay flats, barley or hay or fruit trees or vegetables, flourished in the high water tables and virgin soil. Long before the Gold Rush, the

San Francisco mission and presidio grew food and fodder on farms in present-day Richmond. Promotional books in the late 1800s touted Oakland's inexhaustible fertility to would-be residents. Vineyards thrived in the Fremont area, and Alameda County once led the state in wine production. Had not the vines fallen victim to the phylloxera louse at just the wrong time, the East Bay today might still be a major winegrowing region.

Since then, development has paved the coastal plain and covered up its creeks. Land reclamation obliterated the marshes and cut off the flats from the shore. That's the human history of the flats. Not much geology is left to see.

The Oakland flats are a broad alluvial plain, built of sediment brought across the Hayward Fault from the high hills and laid down in countless thin sheets whenever winter floods burst over the streambanks. Boreholes show that the plain is an intricately layered structure of mostly clay and sand. Buried lobes of coarser gravel mark where stream channels once ran, their courses changing from century to century with the ups and downs of rainfall and earthquake activity. Excavations in the flats have unearthed fossils of extinct ice-age mammals; for instance, in 1964 the Coliseum site in East Oakland yielded bones of mastodons and giant ground sloths.

The structure beneath the surface is the same both onshore and offshore. The San Francisco Bay gradually sinks as mountains rise to its east and west, forming a basin that's continually topped up with sediment. Basins filling with sediment are the geologic setting that makes common rocks like sandstone and shale. Many geologic basins make monotonous rocks, but the San Francisco Bay basin is a complex set of sedimentary formations that hold important supplies of fresh groundwater.

To understand why, recall that ice-age cycles have lowered and raised the sea many times over the last few million years. The Bay is unusual, right now in deep time, in being full of seawater. Its default state, geologically speaking, is a dry valley.

Now picture what this means over dozens of ice ages. When the sea is low, alluvium spreads across the whole basin, some of it coarse and some fine-grained. When the sea is high, marine clay from the Bay blankets the alluvium. The clay spreads inland beyond today's coast during the warmest phases, such as the Sangamon interglacial when marine terraces topped with clay were built around Lake Merritt. So the basin of the Bay is built up in successive layers. Its structure is a kind of layer cake, with sand and gravel standing for the cake and clay layers for the frosting—except all the layers are irregular wedges and tongues and blobs, because the analogies geologists use are always too simple.

There are other irregularities. The Hayward Fault disrupts the streams in the Oakland Hills, which affects their aprons of alluvium downstream. When the sea is low, the streams erode and cut into the older layers, as they did at Lake Merritt. The ice-age canyon of Merritt Creek was one of many like it in the East Bay.

Lastly, the Bay has had different shapes in the recent geologic past. For instance, the Golden Gate did not exist before the Sangamon interglacial. Instead, the river draining the Central Valley flowed down the east side of the Bay past present-day Oakland and entered the Pacific through a valley a dozen miles south of the Golden Gate that geologists call the Colma Gap. Later, transcurrent movement on the San Andreas Fault closed the Colma Gap and opened the Golden Gate. Still earlier, before about 650,000 years ago, the Central Valley didn't drain to the Pacific at all but was instead an enormous freshwater body that geologists have named Lake Clyde. And so on.

But the simple part of the story is how the layers of the alluvial cake handle groundwater. Sand and gravel layers form aquifers, reservoir zones that can store and yield water like sponges. Clay layers, known as confining beds, are basically waterproof. These two types of layers define natural underground spaces that we can use as water resources.

The flats and the dunefields and the shore were generous natural sources of drinking water for the cities that grew up on them. Good water was their last great gift, and one that Oaklanders, true to form, exploited nearly to exhaustion.

Oakland's natural water is harvested by the high hills, which lift the wet sea wind and wring its moisture out in cool rain that otherwise would not fall in our arid climate. In the summer, trees and grass in the hills—especially the ridgetop redwood groves—sift water droplets directly from the fog.

Streams from the hills trickle into aquifers in the flats that wells can tap. Water from wells is the same thing as spring water. Mineral springs aside, groundwater is usually cleaner than surface water from lakes and reservoirs in which, as we all know, algae grows and fish do fish things.

Water supplies concerned Oakland leaders for the city's first eighty years, and the search for new sources visited every part of town. In the original downtown, a shallow aquifer served individual residents and businesses. Elsewhere farmers, for their part, depended on rainfall or high water tables for their crops; thus the outlying foothills around the young city were hayfields, and the floodplain of Fruitvale earned its name as the area's first nursery and orchard district.

When larger enterprises needed water in larger amounts, many located themselves on land with a good well. Oakland's nineteenth-century brewers, for example, tended to migrate to West Oakland. Private companies arose to serve wider areas with water from the largest sources they could find. A few built reservoirs in the heights to capture surface water, as I mentioned earlier, but most providers relied on lowland aquifers. The best of the early wells, in North Oakland, yielded as much as two million gallons a day from an aquifer two hundred feet deep. Within a few decades most of this resource was pumped dry, ruined by pollution, or rendered useless by saltwater intrusion.

As business grew and older wells gave out, Oakland's water companies shifted south to the flats in the countryside near San Leandro Bay, where their wells produced mostly decent water from 1879 until the late 1920s. Water people had the idea that a "subterranean river" existed on the rim of San Leandro Bay. In fact, the aquifers along the shore were a complex of pools fed by individual creeks. Several different companies operated wellfields that produced up to three million gallons per day, enough for a small city but not a large one. In the Fitchburg wellfield, where the Coliseum was later built, the ground was so low that levees were built to protect the wells from storm waves at high tide. But the companies competed themselves into bankruptcy and mergers in the face of repeated droughts, corporate skullduggery, overpumping, unmetered customers who soaked their yards all night, and city governments, especially in Oakland, that imposed ruinously low water rates to keep the voters happy. Even after the providers merged into one after the water wars of the 1890s, the public was poorly served, and every drought exposed the system's weakness. Dams in the East Bay hills were the next step, but money was scarce and water even scarcer, especially as World War I brought new demands from industries.

New laws on the books changed the facts on the ground. A 1921 state law allowing public utilities to serve multiple counties soon led to the creation of a regional water authority, the East Bay Municipal Utility District, affectionately known as East Bay MUD. In 1929, after acquiring the existing water companies from Richmond to Hayward, the new agency filled its mains with pure, fresh water from the Mokelumne River watershed in the Sierra Nevada. The following year it closed all the local wells. Since then, so far, shrinking Sierra snowpacks have met our growing demand.

In aboriginal Oakland, this arid land had water enough for the forests and wildlife and Indigenous tribes. The aquifers protected the streams and their inhabitants from droughts. Frontier American capitalism led to heedless overuse of the pristine resource, while the absence of modern sewage treatment led to widespread pollution. Oakland wasn't the only culprit; cities have degraded the local water everywhere in California. No resource is inexhaustible. Today we still live in an arid land, but pipe surface water from distant mountain slopes. The great Sacramento–San Joaquin Delta was the loser in that deal.

We can make land, but we can't build aquifers, only find them and treat them well or poorly. Despite damage from salt water, pollutants, military wastes, leaky boreholes and such, some of Oakland's aquifers may have space enough for a second life as reservoirs. Groundwater, besides being fish- and algae-free, doesn't evaporate like the water in surface reservoirs. East Bay MUD has begun storing water in aquifers south of San Leandro Bay. Luckily these didn't suffer from permanent land subsidence due to overpumping, a serious problem in California. And maybe new aquifers, never tapped and yet unspoiled, can be reached from the made land around the harbor and airport.

The flats and the shore, and their underground structure of basins and buried canyons, and the rise and fall of sea level are all parts of a bigger story: what happens to all the sediment shed from the continents. The deep seafloor, not the shallow coastal zone, is generally where sediment eventually comes to rest and turns to hard stone. The evidence in sedimentary rocks shows plainly that they form in basins under the sea, yet exactly how they end up on land, even in the highest mountains, was long a stubborn problem for geologists. The solution came when the deep sea was opened to science in the twentieth century, giving rise to the theory of plate tectonics. We learned that not only do rocks form in and return to the sea, but the Earth's very crust is born and dies under the ocean.

It may seem like a paradox when I say this, but it's important to keep the underwater world in mind as I move into the heights where Oakland's rocks live. Most of those rocks were born in a deep-sea realm.

Before leaving the flats, I want to celebrate them: their topography, simple as a blank page and free of bedrock barriers, is what suits them so well for human occupation. The living environment we have built on the flats echoes the complex geologic structure beneath it. When I take a random walk through any part of the flats, I enjoy the ways the layers of Oakland's human history, embodied in buildings and street patterns and vegetation of all vintages, sit in companionable contact. And from everywhere below, the Oakland Hills beckon in the changing Bay light.

MIOCENE EAST BAY

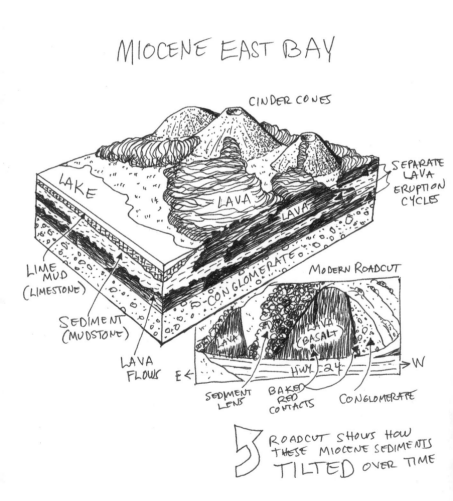

CINDER CONES

SEPARATE LAVA ERUPTION CYCLES

LAKE

LAVA

LAVA

LIME MUD (LIMESTONE)

SEDIMENT (MUDSTONE)

LAVA FLOWS

CONGLOMERATE

MODERN ROADCUT

LAVA

LAVA (BASALT)

CONGLOMERATE

E ← HWY 24 → W

SEDIMENT LENS

BAKED RED CONTACTS

CONGLOMERATE

ROADCUT SHOWS HOW THESE MIOCENE SEDIMENTS TILTED OVER TIME

SIBLEY VOLCANIC
REGIONAL PRESERVE

On a hazy October weekday, this rustic regional park on the far side of the Oakland Hills is as deserted a place as can be found. A steep-walled valley mingles oaks with cattle pasture, and hills stand tall all around, dark with trees or brown with spent grass. A hawk loops overhead on eddies of wind, its trails invisible curlicues; jays in the trees announce my arrival.

The Sibley Volcanic Regional Preserve is in a band of high land a couple miles across, separating the hilly bourgeois suburbs of Montclair, Moraga and Orinda. The volcanic part is in the preserve's northern sector, a no-man's-land of abandoned rock quarries. The southern portion, where the hawk soars and I stand, is abandoned farmland in a side valley off the wooded canyon of San Leandro Creek, upstream from the funky village of Canyon. Marks of its previous owners are strong: dirt roads slashing the slopes, a graded level workyard, shrubs pushing through broken concrete, the fenced-off stream on the valley floor buried in a collapsing culvert.

There are many different ways to enjoy this park. My way is to peer at the land, both at it and into it, peeling away the surficial parts. The things that catch my awareness are half hidden, envisioned as much as seen, in the ground below and in the slopes above.

The trail circling this valley takes ranch roads up the breezy western flank of Gudde Ridge, along the top and down again. A hike on it visits the closest thing we have to the original naked backbone land the Indigenous tribes knew. Sibley is beguiling for that reason, but I'm also beguiled by the glimpse it offers of the truly ancient stories in the rocks beneath. Here, unlike elsewhere, the kinds of clues geologists use to decipher the Oakland Hills are plain to see. Consider this chapter a field trip to both parts of the preserve, with trailside stops and arm-waving.

A few steps counterclockwise up the loop trail is a large gravel workyard with a steel barn. I'm drawn by a clutch of chair-sized boulders near the barn, an appetizer tray brought here from all around the preserve. Their colors and textures testify to a mixture of origins. Black stones, chased with irregular veins of milky white, are former lava flows; red ones pocked with bubble holes are flows that hissed to a stop in freshwater lakes; white ones are limestone from those same waters, as I can verify with acid from my eyedropper by its telltale fizzing.

A little higher on the trail, past a cattle gate, a knob of rock peers down from a cluster of oaks. Broken bits have tumbled to the roadway. Many of the bits are whole pebbles, the size of eggs and buckeye conkers, polished like river rocks but stained with age. I can't resist hefting one in my hand. The outcrop in the oaks is studded full as a fruitcake with these pebbles, a thousand-eyed sentry. It's a type of rock made of stones encased in a hard matrix of fine-grained sand: conglomerate. To me this knob has a familiar face: it's one of a widespread set of outcrops in these hills, known to geologists as the Orinda Formation.

Conglomerate looks like concrete made by someone who's only read about it. This lack of artifice marks it as a natural product. Conglomerate was not meant to be concrete; nothing meant it to be anything. It's the petrified twin of a modern gravel bed at the bottom of the sea. It represents a gravel-bed environment that existed in the geologic past. It tells me a geological short story, a cycle of events that began long before this October afternoon.

Geological stories are of two kinds. One is about the modern landscape: how it was made, or how it's maintained, by natural forces. I call them latter-day tales, and they happened in the last million years or so. During those times, the landscape looked much like it does today. Latter-day tales are what I've been telling so far in this book.

The other kind of story goes beneath the landscape into the deeper geologic past. They're geohistorical tales, stories the rocks themselves whisper to us about vanished worlds and tectonic shenanigans. They explain how larger chunks of California have been shaped and arranged. The conglomerate of the Orinda Formation is part of Oakland's youngest geohistorical tale, which began around ten million years ago late in the Miocene period.

• • •

The rounded cobble I've picked up was once part of solid bedrock in a rugged highland. Erosion—gravity plus gentle weathering in the sun and air—gnawed the bedrock into fragments and sent them clattering downhill, where running water took them up and washed the dirt off this newborn stone. Mountain streams roughly shaped it, click by clack against other stones; larger rivers rounded and polished it in lapidary leisure as occasional floods carried it in stages to the shore. For a while the cobble lay near the coast fondled by waves, then an underwater landslide captured it and carried it swiftly offshore.

Long it languished amid deepening layers of sandy gravel, and imperceptibly the sediment pile lithified into conglomerate stone. After a wait of millions of years, tectonic movements lifted the whole ensemble and tilted it into the air and the light and the weather, part of a new highland in the East Bay hills. A second round of erosion freed the pebble from the hillside just as I came by. The End. That's a short story—erosion, deposition, lithification, tectonism, erosion—that traces the loop of a geologic cycle.

This fossil pebble, preserved from Miocene times, is a sample of an ancient mountain in an ancient landscape. One of the things I love about geology is that it retrieves lost lands. This inarticulate pebble shows directly what kind of rock the Miocene highland was made of; indirectly it hints at mountains before those mountains. Now, for an instant, it warms to the sun and the touch of a living thing. One could say it went to sleep in one world and awoke in another.

And geology's chain of events goes on. Once I drop it, the pebble will roll downhill some more and enter San Leandro Creek, joining the latter-day chain of sedimentary events I described in Indian Gulch and by the bayshore. The pebble might reach the sea, after a thousand years or so, and join a new generation of rocks, but more likely it will wear down along the way and be ground into grit, decay into clay and dissolve into chemical material. Either way, it has entered a new cycle, and a new loop has begun.

Cycles: they're comforting, the root of rhythm, the base of order. We look for them wherever we can. Cycles are the best way we know to gauge the future, to see ahead. Geological cycles like the one the pebble describes are powerful tools for gauging the past, seeing back, probing deep time. Not just pebbles: grains of sand go through this cycle repeatedly. Some of them tell of times not millions, but billions of years ago.

A little beyond the conglomerate outcrop, the trail enters oak-bay woods and emerges to an open sky. Here is the first of many expansive views of the preserve and its surroundings. A picnic table invites a contemplative sit.

Now at the end of the dry season, the grasses are straw, the soil deeply cracked. The days are shortening, and the countryside is ready for the reviving rains. With the last leaf fallen, the brown land pleads: bare earth, wear green. Seasonal changes veil the world with flux. We ponder what lies beyond yearly cycles, beyond mere lives from dust to dust, beyond history. Does geology offer some different perspective as we see deep into time?

Everything human erodes with time: sensory impressions, memories, keepsakes, documents. We differ from other living creatures only in our knack of passing symbolic records, like this pebble of a book, to posterity's fallible hands. Whole landscapes, like the ancient Miocene highland, erode to rubble and are buried. Earth itself seems like a vast cemetery whose monuments are stones. But just as cemetery monuments whisper to us of life, rocks hint at how this planet stays so active, so alive.

The conglomerate of the Orinda Formation is a simple kind of rock, one made of successive layers of sediment. It teaches the basic rules of sedimentary rocks. The first rule is this: they look like the beds of sediment we see around us today, because that's what they used to be—or, stated as a maxim, the present is the key to the past. The second rule is that the layers of these rocks accumulate from the bottom up, their strata recording the passage of time vertically as surely as the rings of tree trunks, counting from the heartwood out. Upward is younger and downward is older.

These rules were among several principles stated by Nicolaus Steno, the founder of geology, in 1669. They still teach them in geology school, plus all the exceptions we've learned since then.

Steno had to argue carefully and at length to persuade his peers of truths the ancients had never known, realized by analyzing things instead of words. This was the new kind of knowledge Francis Bacon had called for a generation earlier. Before that, pebbles in conglomerate aroused little curiosity. The ancient Greek and Roman authorities had said nothing about them, and for those who wondered, there was one response that dismissed all questions and trivialized all wonderment: the Bible says God made the world a few thousand years ago, and whatever we see was just made that way.

The biblical world is not a cyclic world, like the Hindu cosmos, but a dramatic one: it begins with a word and ends with the Second Coming, as surely as a play on a stage. The biblical Earth is purpose-built, a set for one performance with its mountains ready-made. But conglomerate, a petrified gravel bed, prompts a question that led thinkers to a different worldview: Once these pebbles were laid low, what raised them high again? Anyone can see how mountains are eroded, but how are they built?

Today we know that the Earth constantly renews its mountains—it moun-tains itself—in a system of repeated cycles. This insight was first articulated in the late 1700s by a Scottish gentleman farmer named James Hutton. He held that the Earth was designed to sustain creation indefinitely, for as much time as the biblical drama needed, by steadily producing soil for farmers. The Earth system, Hutton argued, is "a machine of a peculiar construction," a self-regenerating stage set that takes countless years, truly deep time, to serve God's design: "We perceive a fabric, erected in wisdom, to obtain a purpose worthy of the power that is apparent in the production of it."

Stripped of its Christian scaffolding, this vision of the Earth machine is still a wonderful thing. In geology it's brought us far, although Hutton's favored mechanism, in which land and sea switch places like paired pistons by the action of subterranean heat, has been superseded. The modern theory of plate tectonics explains all of Oakland's rocks, including the ones in Sibley Volcanic Regional Preserve.

As for my question, whether geology helps us see something beyond another set of cycles, Hutton's famous last sentence was noncommittal: "The result, therefore, of our present enquiry is, that we find no vestige of a beginning—no prospect of an end." And though geologists no longer accept his answer—we *do* have a big story—it's good enough for my purposes. The Earth does preserve rare vestiges of its beginning, but not here at Sibley or anywhere else near Oakland.

The upper valley of south Sibley rises past my picnic table. Its rocky slopes form saggy, curvilinear surfaces like blankets draping a raised knee. The sedimentary rock of the Orinda Formation, rugged as it looks, is not strong stuff. Here are none of the lumpy hillsides and bedrock knobs of the Piedmont block; rainfall and gravity carve the Orinda Formation cleanly, without undue resistance. It crops out in an evenly layered sequence thousands of feet thick, encompassing almost the whole slope. The higher goes the trail, the younger grow the rocks. The rock strata, alternating beds of coarse and fine material, form lines across the valley walls.

The ascending trail passes many roadcuts, the geologist's best friend. They expose the bedrock in detail. The notable thing they reveal is that the rock layers are all tilted steeply downward into the hillside. Another of Steno's rules of rocks is that strata are born horizontal; if they tilt, something tilted them.

Also on the trail are small landslides. Study these and more can be recognized as pockmarks and scoop shapes on the slopes. To geologists, landslides are nature's efficient way of moving matter downhill in large masses. In the million-year movie that geologists see wherever they look, individual landslides disappear and the rising hills melt like ice cream in the sun.

• • •

The top of the valley, where the trail turns north along Gudde Ridge, is a good place to try the mental exercise I've mentioned before: to look at this landscape and picture its changing appearance over Oakland's history.

This was a vibrant neighborhood when Europeans first came into the country. The tribes came to pick mushrooms in winter, tend useful plants with fire, harvest game animals in the meadow and catch salmon in the creek. When wildflowers bathed these hills in color every spring and summer, the tribes knew the virtues of every species, but surely too they were pleased and awed by the kaleidoscopic bloom.

Don Joaquín Moraga, who was granted this land in 1835, employed Indigenous herders who watched the wildflowers fade and fall prey to foreign grasses.

Next the McCosker family and their neighbors used this valley for a few generations, mainly for cattle ranching. As a side hustle in the mid-1900s the owners opened a small quarry, now overgrown, to produce crushed rock for roadways and local builders. They used the quarry waste to bury the stream and build flat workyards. When the McCosker descendants cashed out, the worn-out valley became mere property, an abstraction dealt between syndicates, a card swapped among players in search of good hands. The regional park district took up this parcel in 2010 to add to the Sibley preserve. The district recognized an asset in its varied wildlife habitats, and today I recognize another in its geology.

South Sibley is a wildland park, a contradiction in terms. The park district calls it a "natural unit": a zoo for land. Visitors are to stay on the paths and keep their hands off. Its keepers tend this caged countryside guided by the terms of their legal charter. They're unburying the stream and plan to prod it, like a stunned victim, back to life under the name of Alder Creek.

If the tribes were able to resume their seasonal burning and other traditional practices, they could start to bring the hillsides to something like their pre-contact state. If so, it would mean acknowledging that what we call wildness was something that people arranged. Indigenous land was not wild land, but home land. What our laws define as wilderness, an area "untrammeled by man," is depopulated land.

For the moment, this part of Sibley preserves degraded habitat, pasture left as the McCoskers homesteaded it in the 1860s, a diorama exhibit of nineteenth-century ranchland. Although long-range plans include controlled burns, the park planners have hesitated to set dates and run trials. Meanwhile natural succession advances as the grasses give way to scrub and forest, laced with deer trails. Oak-bay woods are gradually covering the rocks.

Now let's look a little farther back, to the dawn of human presence. Since the last ice age reached its maximum, some twenty thousand years ago, these hills have changed costume several times. At first they wore cool green savannah, then temperate-zone forest and chaparral as the ice age faded and the first humans arrived via northern Asia. As century followed century and the climate warmed and dried, these tribal cultures learned to optimize the grasslands ecosystem, and themselves, for mutual well-being. The coat-of-many-colors the tribes perfected was perhaps the hills' finest outfit.

To the geologist, the changes that swept this landscape since the Last Glacial Maximum are as fleeting as the shadows of clouds.

Human time is pitifully short. Geologic time is dizzyingly different. It's not so hard to picture a thousand years, one millennium: about fifteen life spans of threescore-and-ten. Now take that millennium and repeat it a thousand times in a row: a million years. Then take that million years and repeat it too, four-and-a-half thousand times. The Earth's age is 4.55 billion-with-a-B years. Time of this caliber is inexhaustible, a colossal stage on which we are but flickers of activity, little more than those hawk trails in the air.

To geologists, a million years is not quite a snap of the fingers; more like the tick of a clock. Gaze through so much time and even the rarest events—thousand-year floods, great earthquakes, century-long droughts—are part of a continuum. Clashes of continents, oscillations of ice ages and monster impacts from space are part of the mix of rare events. In the geologic present they, not everyday events, are what really shape the world.

In the finger-snap of time since the Last Glacial Maximum, beneath its changing costumes the land has looked the same. Geologists, with their million-year vision, watch water and gravity cut the land down and see it rising up for more. This world consists of remnants of different, much earlier ones; the pebbles in the conglomerate have told us so. Now we're getting somewhere.

The view from the crest of Gudde Ridge is wide and fine. From here this ridge is seen as one of several, aligned like a pod of whales along the grain of the Coast Range, northwest–southeast. Just up the ridge, to the northwest, rises Round Top, the most prominent summit on the Oakland skyline. It's also part of the Sibley preserve. Fourteen miles due east stands Mount Diablo with its twin peaks, and on the clearest days the Sierra Nevada can be seen, its grand white ranks ten times more distant still. They all are part of the biggest, deepest story of California.

Here up near the crest, the rocks along the trail change from conglomerate and sandstone to spikes and blades of dark lava, known to geologists as the Moraga Formation. This was what the big rock quarries were after in the northern part of Sibley. The virgin outcrops here are daubed with lichens, so thoroughly that the lava itself can barely be seen. Like the rocks below, these lava beds are also steeply tilted into the ground. The ridge's hard cap of lava protects the Orinda Formation beneath it.

Fossils of extinct animals and plants in the Orinda Formation tell us its age and environment. They say these rocks formed in a forested place on the coast, somewhat wetter than today, during the late Miocene period, around ten million years ago. Just north of Sibley, a major freeway passes through the Oakland Hills in the Caldecott Tunnel. When the tunnel's fourth bore was drilled a few years ago, a "paleo" contractor sifted the tailings and recovered thousands of fossils: seashells, bones of extinct land animals ranging from shrews to rhinoceroses, and fossil leaves of the plants they fed upon.

The shells appeared in the lower part of the formation, the land fossils in the upper part. This finding tells us that the environment slowly changed from a shallow sea to an alluvial lowland. It could have resembled today's East Bay—or the San Mateo Peninsula, across the Bay, because traces in the gravel show that the streams flowed east, not west, to the sea. Probably the land was being uplifted.

The pieces of gravel consist of Franciscan rocks, ten times older than the fossils. They match the rocks in a set of hills far to the northwest, on the other side of the Hayward Fault family. The faults broke up this Miocene geography while eruptions of lava, the Moraga Formation, covered the scene. I'll get to that part next.

The trail, turning down from Gudde Ridge and completing its loop, passes thickets of trees that hide the Mc-Cosker family's old quarry. These pits, foundations and staging areas will be prettied up some day for the diorama. The park district will put in picnic spots for visitors, and those people may then lay hands on the rocks, just as I do, and lose their minds in the cycles of deep time.

$$\cdot \; \cdot \; \cdot$$

The north part of Sibley is what makes it an official volcanic preserve. Rock quarries there exposed the guts of a small volcano that last erupted some nine million years ago. The volcano isn't something one can see, not even on the geologic map: it has to be inferred from outcrops and perceived in the mind. The park's brochure and trail signage introduce the evidence, and understanding deepens with repeated visits. The volcano's story is a small local episode in recent geologic history, part of the larger tale of how the Bay Area was assembled.

I've been mentioning various ages as I talk about these rocks. How can we be so precise, and so confident, about them?

Rocks have two kinds of age, relative and absolute. For these concepts I like the analogy of coins. Ancient coins of the Roman empire didn't have dates on them; instead they bore portraits of the emperor who had them minted. Archaeologists can estimate the age of an ancient site from the coins found in a dig, knowing the years the emperors reigned. But imagine we had no other records of ancient Rome. In that case, we could do lots of excavations and gradually work out the correct order of emperors from the coins, but the timeline of emperors—relative ages—would be our only dating tool.

Relative age is what the rocks of the Orinda Formation illustrate: their vertical position in the geologic record and the mix of fossils they bear. Their age, late Miocene, is based on fossils. The Miocene period is officially defined not by a span of years but by a vertical sequence of sedimentary rocks in a specific place with a particular series of fossil organisms.

Relative ages of rocks, worked out over a hundred years of research, were all we had when the twentieth century began. Then we found new ways to determine absolute ages, based on the radioactive decay of uranium or potassium atoms trapped in the mineral crystals of lava as it cooled from its molten state. The crystals, in my analogy, are like modern coins that are stamped with the year they were minted. These days we can determine the absolute ages of suitable rocks with accuracies better than one part in a thousand.

Absolute ages from the lava in the Moraga Formation help us assign dates to the Orinda Formation beneath it. The oldest ages are about 10.2 million years, meaning all of the conglomerate is older. The youngest are about 9.0 million years, so the volcanic field was active for roughly a million years. That's typical for small volcanic fields—just a tick of the clock.

When dating rocks this way, it's important to prove that the lava lies directly on the conglomerate, without a gap in time between them. That proof can be seen north of Sibley in the high roadcuts of Route 24 east of the Caldecott Tunnel, where spectacular exposures show the tilted Orinda and Moraga Formations in perfect contact, lava over gravel. The gravel is baked to extra hardness and the lava flows are turned red, as surely as iron rusts in rain, where water in the underlying gravel flashed into superheated steam. The event is written plain in stone, something for drivers in rush-hour traffic to savor.

In north Sibley, volcano parts are displayed where two rock quarries exposed them. The Ransome Company opened the first one in 1904, employing eighteen men. "The rock is a fine-grained basalt, and is used for macadam and concrete," the State Mining Bureau reported at the time. "Some gutter rocks are sorted out. The rock is hauled to Oakland and Berkeley by wagon." In much of downtown Oakland, the gutters are lined with loaf-sized blocks of dark-gray lava, their tops rounded by a century of street wear. A company in the Kaiser Industries constellation produced crushed rock at the second quarry.

The regional park district acquired this land in pieces starting in 1934. For many years it left the park alone, unscathed by popular attractions, except to add a few campsites and a fire station, to serve the high hills, on Round Top.

The first of the played-out quarries was added in 1977. At the time, no geologist had properly mapped the land. Steve Edwards took on the job as a personal challenge and gave the park its mission. "At the time I was a graduate student in paleobotany at the University of California, Berkeley," he later wrote, "and I decided to make a geologic map of the Preserve to improve my mapping skills." The park soon added "Volcanic" to its name. Throughout his long career as manager of the botanical garden in Tilden Regional Park just up the ridgeline, Edwards was also Sibley's geology maven.

A balance sheet for these quarries, like other quarries in Oakland, would weigh the benefits to the city and the jobs for the workers against the ruination of the land. On the benefit side, I can add the unique lessons this autopsied site offers the public, thanks to Steve Edwards. I also value the evidence of Miocene central California that Sibley yields to science.

All of the rocks in north Sibley have been steeply tilted; volcanic features that formed on level land lean sharply eastward now. The park displays places where lava flowed, where it sprayed in fountains of ash particles, where it baked the ancient soil and where it invaded marshes and shallow lakes. Parts of the underlying Orinda Formation preserve the conduits that fed lava to the volcano. Much of the lava is full of little holes that were once bubbles of gas, mostly water vapor. In some places those fossil bubbles later filled with mineral matter: white opal, bluish chalcedony, buff zeolite or green celadonite.

Besides lava, the park contains beds of mudstone from the times between eruptions, periods that probably lasted thousands of years. It also has fossil-bearing beds of white, porous freshwater limestone that formed in shallow lakes.

The park's eastern slopes, facing Orinda, show much more of the limestone. In the early days this material was mined on a small scale and roasted to make lime, an essential ingredient in mortar.

The way to appreciate these things in the conglomerate and the lava and the limestone is to search for them, marvel at them and leave them be for others to savor. In a region of geological treasures, this volcanic park is Oakland's rarest gem.

The rocks of the Sibley preserve are part of the story of the Hayward Fault. Just as it currently disrupts the stream valleys in the Oakland Hills, the fault has been rearranging the greater East Bay for some twelve million years. The transcurrent faults of the Coast Range were born when a change in the shuffling tectonic plates started in central California near Monterey, about twenty-eight million years ago, and propagated from there to the north and south. Today these faults, of which the San Andreas is the best known, reach from just south of Oregon to the coast of central Mexico.

As these faults reorganized the Earth's crust, pieces of California's western edge rose and fell and drifted generally north, forming large islands and peninsulas in a coastal sea much like the Channel Islands off Southern California. One of these highlands shed the pebbles and sand of the Orinda Formation. Soon afterward, the growing welt of the Coast Range cut off the Central Valley from the sea. The old California broke up and modern California began to form.

The breaking crust allowed hot rock below to rise and start melting, as it does when the pressure on it is released. Lava soon came up in a string of volcanoes, including the Round Top volcanic center. Afterward the Hayward Fault and its sister faults tore off parts of the lava field. One piece is in the hills above Berkeley, just a few kilometers away. Other pieces were moved about sixty miles north to Petaluma, where they're called the Tolay Volcanics. The Orinda Formation's counterpart there is the Petaluma Formation. The scattered pieces bespeak an original unity.

This part of the Oakland Hills is deeply connected to the North Bay in ways that continue to challenge researchers. Not only have slices of the crust been separated like a spilled loaf of bread, but the faults between them have moved at different rates in shifting alliances. Some faults ceased activity and died as others were born to replace them.

More recently, transpression raised the Coast Range, folding and even overturning the rocks that formed here. It also unearthed much older rocks, with far different stories, elsewhere in the Oakland Hills.

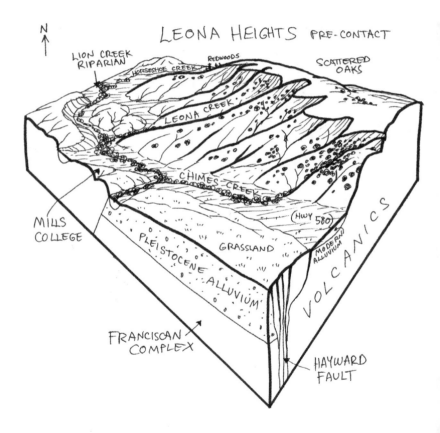

LEONA HEIGHTS AND THE
SOUTHERN OAKLAND HILLS

The Oakland Hills form a great wave overlooking the city, one that subtly varies along its crest with the different rocks that hold it up. Leona Heights, near the south end, is the most forbidding part. Unlike the rest of the hills, where houses cover the slopes like spume on surf, here they're dark with oaks. Where Interstate 580 turns in close, Leona Heights looms tall. Here the wave seems closest to breaking. These hills rise high upon the Hayward Fault. Here more than anywhere else in town, humans have dug and delved, sent mine shafts deep underground and left a mess behind.

Leona Heights looks like its name suggests: mountain-lion country, intricate and inaccessible. I know paths in the southern Oakland Hills where fresh deer bones lie by the trail. Though I've yet to see the secretive cat that leaves them, my senses prickle more than usual in this part of the hills. As happenstance has it, Oakland's oldest rocks are underfoot.

The name comes from the Spanish by a doglegged route. When the original Peralta ranch was divided in 1842, the boundary between Antonio's and Ignacio's shares was the Arroyo de Leon. Today we still call the stream Lion Creek, but "Leona Heights" was a real-estate scheme in the 1890s.

These highlands are defined by a body of rocks that plate tectonics carried here from tropical latitudes during the Jurassic period. Much of the human history of Leona Heights centers around this rock, which early geologists gave the mistaken name Leona Rhyolite and which I call the Leona Volcanics. They crop out in a stretch of fourteen miles from Redwood Road south past the city line all the way into Hayward. A few fragments crop out in North Oakland and Berkeley. The Leona Heights neighborhood, where their presence is strongest, is what geologists might call their type area. The Leona Volcanics were born about 165 million years ago, which is about 3 percent of the Earth's age.

Like most parts of Oakland, Leona Heights yielded a succession of resources derived from its geology, each one exploited to exhaustion with damage left behind. It began with the Ohlones, who used to come up this way to work a large deposit of ocher.

Ocher is a claylike pigment that humans have used since at least the time of Europe's ancient cave paintings. To geologists, it's a spectrum of earthy iron oxide minerals, ranging in color from mustard yellow through brick red to dark brown, that form on suitable rocks by surface weathering. The pre-contact Ohlones traditionally used powdered ocher to make body paint, an essential part of the daily costume in their unclothed culture, and for protection from sun and insects. An unheralded exhibit at Holy Names University, at the north end of Leona Heights, features piano-sized boulders of the dark-red raw material, dented with coconut-sized pits like those the ocher miners made.

Ocher was also a grave good sometimes used to dress the dead. When a reckless reporter for the San Francisco *Call* named J. H. Griffes hired two men in 1892 to shovel a trench across the Alameda shellmound, he noted the presence of "a curious looking red baked clay or adobe" among the crumbling skeletons with no show of recognition.

Leona Heights ocher was traded in a lively regional economy. The Ohlones exchanged it for East Bay sea salt from the Coyote Hills area, cinnabar pigment from the South Bay, abalone beads from the seacoast, obsidian tools from the Napa Valley, pestles or grinding stones from distant sources, and more. Indigenous Californians have not forgotten this ancient trade, and ocher still forms in these hills. The Ohlones were not the ones who exhausted the deposit; Americans did that, as they did with every resource they found.

The first resource the Americans tapped was redwoods in the high hills. The great ridgetop groves extended an arm down into the northern part of Leona Heights, although the rest of this district was originally grassland rather than the oak woods of today. Between 1840 and 1860, Americans felled every tree but one. The last old-growth redwood on this side of the Bay, nicknamed "Old Survivor" and nearly five hundred years old, is in Leona Heights on a cliff in Horseshoe Canyon.

Once the redwoods were gone, water was the next thing found to exploit. The Contra Costa Laundry set up shop in Lion Creek's headwaters in 1859, laying linens out on the grassy slopes to dry and bleach. Although the business washed away in the terrible winter rains of 1861–62, the area was known long afterward as Laundry Farm. Later the water companies sought supplies here, but not with great success. The mines of Leona Heights polluted what water there was.

The East Bay Mexican ranchers weren't miners, but during the Gold Rush years Americans and other outsiders turned over every stone in California. The smart ones learned or absorbed the practical, resource-focused geology of their time and staked claims on all types of useful ores. But in the Oakland Hills the Gold Rush was a bust, and that was it for the next thirty years at Laundry Farm.

Elsewhere in Oakland, there was briefly a gold mine, started in 1864, on A. D. Pryal's ranch next to Berkeley. The ore, described as "a vein of auriferous quartz," was probably in a small body of highly altered material, called silica-carbonate rock, caught up in the Hayward Fault. There were one or two other reports of gold near the fault at the time. Everything else was rock quarries.

Meanwhile, Laundry Farm got its first rail line in 1887, built in hopes of sparking resort development and a residential land rush, but a bank failure ended the scheme. In 1895 the California Railway Company took over the line and built the Leona Heights Hotel and resort on a sunny slope near Horseshoe Canyon. There were trendy "trolley parties" in the fresh air and scenery, with music every Sunday by Homeier's Band. The idea was to subdivide the hillside property and make it a district of country homes. There was another angle involving water rights. None of these plans panned out, but the name stuck. The Leona Heights Hotel burned down in 1907.

Nevertheless, the rail line made a profit serving quarries and pyrite mines in the Leona Volcanics, which I'll get into later. There was blasting every day for decades at Leona Heights. Passenger trains shared the track with loads of ore and crushed rock. Large grass fires swept the area from time to time, and the train line was the scene of several deadly accidents. The residents stuck it out; many of them worked in the mines and quarries. They were mostly European immigrants, paid less than two dollars a day—if they made any more, they might move away.

Meanwhile, other large landholders around the southern Oakland Hills were creating new streetcar subdivisions. Their selling points were healthful air, views, sunshine and land enough to raise chickens as well as children. When the Realty Syndicate consolidated the East Bay's light-rail systems in the early 1900s, real-estate ads touted the easy commute to San Francisco. Not long after, the automobile changed everything and development was favored elsewhere. For decades, the steep areas of the Leona Volcanics and related rocks were left vacant, but slowly they too became residential districts and open-space preserves.

As development gained ground and digging died out, first at the mines and then at the quarries, the seclusion was short lived. Highways came through Leona Heights, then freeways. Today where modest residential streets wind through the scarred hills, the exhaust and tumult of commuter traffic replace the smoke and bustle of the mines and quarries—and wildfire still threatens the steep, untended hillsides.

• • •

Time for a closer look at the rocks, the Leona Volcanics. The pyrite the miners sought and the ocher the natives prized are accidents of geology that are connected. To tell the story of the Leona Volcanics, and how they got their ocher and pyrite, I must set the stage with a dip into plate tectonics, the framework of knowledge that explains volcanic rocks.

Plate tectonics is the Earth machine that keeps the world from wearing out and running down. The planet's outermost shell is broken into sections—large and small, thick and thin—that move around as aimlessly as the scum on a stewpot, about a billion times slower. Those sections are the plates, and *tectonics* refers to their large-scale movements.

The outer shell of plates is a layer of solid rock, cold on top and warm below, called the lithosphere, scientific Latin for "stone zone." Beneath it is the mantle, a hot domain of solid rock nearly two thousand miles thick. Plate tectonics describes how the lithosphere is created from and destroyed in the mantle, renewing the Earth's crust in the process.

The plates can move and the mantle can stir, even though they're solid rock, because to put it as simply as possible, hot rock responds to physical forces exactly *as if* it were thick tar or putty. High pressure from the great weight of overlying rock keeps the mantle from melting. But rocks come close to melting near the top of the mantle, and the plates ride upon that softer zone.

In the first chapter, I made the analogy of a soccer ball whose polygons can be moved like a Rubik's Cube. Plates on a sphere mainly interact in three basic ways: they move toward each other, move apart, or slip sideways past each other. As geologists put it, the motion between plates is convergent, divergent or transcurrent. The Hayward Fault belongs to a transcurrent plate boundary, as I've mentioned, but when the Leona Volcanics were made, California was part of a convergent system.

Convergent zones are the power source for plate tectonics. Let me sketch the basics.

The lithosphere comes in two types, with different thicknesses: oceanic lithosphere is about ten miles thick and continental lithosphere about a hundred. Oceanic lithosphere forms where plates are spreading apart. It's always made of lava, mainly the heavy black type known as basalt, so it lies low and is always covered by the ocean, hence the name. The older it gets the colder it gets, and as it cools it shrinks. This raises its density, after a few tens of millions of years, to the point where it's gravitationally unstable, denser than the mantle beneath. Like a sheet of paper blown into a pool, oceanic lithosphere stays where it is until one edge begins to sink into the mantle. Plates are pulled around by their sinking edges, and that force—gravity working on rocks of different densities—is what drives the Earth machine.

The continents, on the other hand, are made mainly of granite, a lighter-colored, less-dense rock that like the scum on the stewpot never becomes dense enough to sink. Continental lithosphere rides high. Continents move wherever the plates take them, breaking up at divergent zones and being pushed together at convergent zones.

As I said, the lithospheric plates are made of solid rock. Now I can zoom in to molten rock: lava and volcanoes. And speaking like a geologist, I'll refer to melted rock as *magma* when it's underground and *lava* when it's erupting and after it cools to solid rock.

When rock melts, it expands, which means that magma is less dense than the rock it melts from. So magma rises by simple buoyancy, making its way upward through cracks in the overlying rocks.

Rocks in the lithosphere produce magma in two ways: they're *allowed* to start melting when the pressure on them decreases, and they're *encouraged* to start melting when water is added.

The first occurs in divergent zones, where pressure is released and magma forms at seafloor spreading ridges. Oceanic plates grow along their trailing edges as the rising magma freezes into lava. That's how the Leona Volcanics started out.

The second occurs in convergent zones, where cold oceanic plates sink beneath other plates in the process of subduction. I say more about subduction in the next chapter, but basically subducting plates carry water down with them. This water is released under the growing heat and pressure, then rises into the overlying plate and triggers melting there. Chains of volcanoes form above subducting plates.

To sum up, the ocean floors, like conveyor belts, continually sweep into the mantle at subduction zones, where they release water that bastes and stirs the continents with magma and volcanoes. That's the elegant machine that keeps the Earth's surface fresh. Now I can tell how the Leona Volcanics came to be and how they came to support a generation of mines in Oakland.

Geologists have published several different stories that account for the Leona Volcanics. None of these hypotheses may be entirely true, but each one is persuasively told by experienced researchers. Each story favors some lines of evidence and disparages others. All of them show, though, that the early Bay Area geologists were wrong to call these rocks "Leona Rhyolite." That is a stiff type of lava that flows from volcanoes more or less intact, whereas the stuff of Leona Heights began as volcanic ash.

The following version was published in 2008 in a paper by Clifford Hopson, a distinguished researcher at UC Santa Barbara, and three coauthors. I've picked it because it's simple to tell. Once upon a time, during the middle Jurassic period, a new piece of ocean floor—let's call it Leonia—formed where two plates were pulling apart, somewhere in the paleo-ocean Panthalassa, what's now called the Pacific, in tropical waters near the equator. Its birth was accompanied by vigorous eruptions that built a long, thick pile of lava and ash. Leonia found itself riding east and north on its oceanic plate toward, and under, the paleo-continent Laurasia in the sector that later became North America. Along the way, about fifteen million years later, another spreading zone zipped open beneath Leonia, injecting it with fresh magma that juiced and reheated the volcanic pile.

When Leonia reached the subduction zone at Laurasia's edge, it was briefly smeared along the edge and disassembled, then sliced off of its parent plate when a new subduction zone formed behind it, an orphan handed off to a new parent. Now part of North America, it was gradually buried deep in sediment washing off the continental mountains. Leonia was reawakened and further disrupted in Miocene time when subduction changed to the transcurrent tectonics of today, and erosion exposed its warped and folded pieces in the rising Oakland Hills. The End.

In every version of this story, the essential part is that the Leona Volcanics formed on the seafloor on a piece of oceanic lithosphere that later, instead of being subducted, stuck onto the continent. This uncommon feature—a chunk of seafloor crust marooned on land—is called an ophiolite. Having pieces of one in Oakland is a privilege.

Another piece of this ophiolite is the gabbro of San Leandro, the body that hip-checked the Piedmont block as it rode along the Hayward Fault. Gabbro comes from deep in oceanic plates, where slow cooling allows its minerals to grow into large, visible grains. Our gabbro is a light-colored variety with a slight bluish cast caused by chemical reactions during its long cooling process. It lies next to the volcanics south of Leona Heights, but the two rock bodies aren't directly related; they just ended up neighbors. The

Hayward Fault is breaking chips off the gabbro and bringing them north, miniature versions of the Piedmont block. One of them supported a quarry at the head of 98th Avenue that's now the home of Bishop O'Dowd High School and a large steel-capped reservoir next door. A third piece of the ophiolite is exposed along the ridgetop: a large body of serpentine rock.

Now on to the volcanics themselves. The fine-grained ash formed as lava erupts in cold seawater is mostly rock glass, unorganized mineral stuff that has cooled too quickly to sort itself into proper mineral grains like the kind in granite or gabbro. Volcanic ash is chemically vulnerable, ready to lose its glassy nature given an opportunity. That's what happened to Leonia as hot-spring activity on the seafloor suffused its pile of ash and lava with a mix of superheated seawater and mineralizing magma juices. This hydrothermal process was repeated every time Leonia was injected with more magma.

Thus the Leona Volcanics turned from black ash into hard, light-colored rock shot with new minerals and intrusions of dark lava. That's when they acquired the abundant supply of pyrite, mostly sprinkled in tiny grains, that supported the mines of Leona Heights.

• • •

Pyrite is a pure compound of iron and sulfur, a mineral whose brassy, metallic crystals form shapes like those in Cubist art and give it the nickname fool's gold. It gets its name, "fire-stone" in scientific Latin, from a tendency to spark when struck.

Like most minerals, pyrite is happy underground where air won't reach it, but uncomfortable at Earth's surface. Water and air slowly break it down into sulfur and iron compounds, and several different classes of microbes eat it as part of their life chemistry. At Leona Heights, the sulfur gradually turns to sulfuric acid and flushes away with the rain and groundwater. The iron remains as various rusty oxides.

The Leona Volcanics are so impregnated with pyrite that in most places, iron oxides quickly paint the fresh rock in colors from honey through bright red to dark chocolate. Do what geologists do and look for fresh rock in a stream-bed. There the well-scrubbed stones are nearly white, with a dull finish, and are very hard. They have no strata. Some contain mottles of blue-green celadonite, created by chemical alteration, and bits of glassy black material that escaped alteration. Outcrops often display slickensides, the polished surfaces like stretch marks that are made by small faults.

The massive pyrite lode that fed the mines was originally covered by a blanket of iron oxides—the ocher deposit that so interested the Ohlones. It must have accumulated over tens of thousands of years.

Then a capitalist found it. That man was Fritz Boehmer, a restless Prussian immigrant who did well in the Gold Rush and then so prospered in business that he became a founding father of the city of Alameda. In the late 1880s he purchased ninety-two acres just north of Laundry Farm along the old road to the redwoods, where the ancient ocher deposit lay. He mined the natural pigment and built a factory to manufacture what was later described as "a good article of domestic paint," but after a few years the factory burned down.

Boehmer improved his property over the years, re-grading Redwood Road at his own expense to make the old "blue road," named for the serpentine rock it exposed, more tractable. Then he opened a hotel and store. He had a keen interest in the natural springs on his land. Having heard an old-timer's story about Lion Creek, which produced a short-lived surge after a large earthquake in 1838, Boehmer was sure the area held lots of water hidden underground. Eventually, around 1910, the People's Water Company did construct a small reservoir near his land, so he was not entirely wrong.

In 1898 he passed around a rock sample from Leona Heights, laced with yellow metallic veins he claimed were gold. Newspaper accounts were gently skeptical, and so was he: "I do not personally care to develop any mining property, as I never have any more luck. . . . I am more interested in mining for springs." It seems to have been a false alarm, or perhaps another rare specimen of gold quartz related to the Hayward Fault. But in the summer of 1900, while digging the foundation for a new hotel after his first one burned down, he struck pyrite pay dirt.

Pyrite mining was the chemical industry's chief source of sulfur at the time. The crushed ore was roasted in air to oxidize the pyrite, yielding gaseous sulfur oxides and solid iron oxides. This process ignites the mineral and produces more energy than it absorbs. The gas was collected and refined, and the solid was waste or a by-product passed on to other purposes.

That year Boehmer started the Alma Mine, named for one of his daughters. He planned to harvest valuable impurities—gold, silver and copper—from the roasted pyrite residue. The old Forty-Niner's heart was set on a gold mine after all.

Shortly thereafter, the Leona Mine opened a rich pyrite deposit a short distance southeast. A few years later, the Leona Heights Mine opened high in a small gulch a mile or so southeast. Although the gold and silver yield never paid, the Alma and Leona Mines together produced about 160 tons of copper during their active life. The Stauffer Chemical Company sent the output of the mines down the streetcar line to a refinery near the shoreline at the Melrose station, where it manufactured sulfuric acid, then as now a basic necessity in industrial chemistry.

An expert from UC Berkeley may have been the first to recognize the mines and the ocher beds as parts of something larger: an ore system consisting of an "iron cap," or gossan, that overlies a sulfide deposit. Whoever that person was, word got around, and a brilliant Berkeley student named Waldemar Schaller brought Boehmer and his Alma Mine into the scientific literature. An Oakland native still living with his north German immigrant parents, young Wally Schaller must have bonded with the Prussian capitalist in their mother tongue.

In 1903, still an undergraduate, he published a substantial paper based on some of Boehmer's favorite specimens, describing red and yellow ochers as well as a dozen different minerals that combined iron, copper and sulfur. The secondary minerals around the edges of the deposit particularly excited him: "These natural vitriols occur in such good crystals that an interesting crystallographic study can be made of them." One mineral was new to science, a delicate sky-blue hydrated copper sulfate that Schaller named boothite to honor Edward Booth of the Berkeley chemistry department. Boothite was the first of forty-one minerals Schaller discovered and named in his celebrated career.

The mines were problematic. When pyrite dust burst into spontaneous flame and triggered underground fires, Leona Heights had a chronic pall of caustic smoke. The mine operators could only seal off the burning area and wait for it to snuff out. Afterward, the area had to be inspected, the timbers replaced and the mine walls bolstered. If other parts of the mine stayed open, production could stagger on. But when cheaper sources of sulfur near the Gulf of Mexico wiped out the pyrite market, Oakland's last mine closed in the 1930s, and the abandoned openings were gradually sealed. Old-timers tell of youthful exploits in the crumbling tunnels, and ruins lurk in the abandoned tracks and overgrown woods of the heights.

• • •

Not much is left in Leona Heights for rockhounds, but geologic activity continues. Although mining, road-building and residential development have left scarcely a trace of the ocher deposit and the iron cap it represented, roadcuts are accumulating fresh coats of ocher. So are the walls of the former Crusher Quarry, overlooking a tract of 1950s-era homes that replaced miners' bunkhouses. Some of the ocher looks pure enough for the Ohlones to collect again, or for an artist to make real sienna and proper umber pigments from scratch. Elsewhere in the Leona Volcanics, acid alteration has turned pockets of the rock into kaolinite, or china clay. None of these materials form economic deposits—not even little specimens worth selling in a rock shop—but they show that the Earth is constantly making them in everyday places. They aren't valuable, just interesting.

The rock quarries have left the largest marks in Leona Heights. The great scar of the Leona Quarry, visible from across the Bay, is the product of digging that began in 1904 and ended, after several changes of ownership, in 2003. It was the last active quarry in town. Lately it has slowly filled with close-packed townhomes, under beetling slopes equipped with terraces and berms to stop runaway boulders. At the same time, high-end residences along the ridge above the scar have fenced off the magnificent Bay views from the public.

A short distance north, the hilltop site of the old Leona Heights Quarry, opened by the Realty Syndicate in 1896, became the home of Merritt College in 1971, and the views from there, still open to the public, are wonderful. Odd stony rises punctuate the campus, much like those at the Sibley preserve. An undeveloped area on its north side once held a deep pit, inviting mischief, that decades of local residents called Devil's Punchbowl.

The stream draining the site, Horseshoe Creek, runs through a steep gulch preserved in the city's Leona Heights Park. Unlike other Oakland canyons, its creekbed is lined with bare boulders, some the size of trucks. My theory is that the surprising ruggedness of this little canyon results from acidic water, both natural flow and groundwater released from the quarry. As the water altered the rocks to clay and washed it away, the bedrock broke into large chunks that were left dry as the stream rapidly cut down past them.

The pyrite mines also have left marks from their decades of activity. Although the mouth of the Alma Mine is now sealed and covered by a tight cluster of houses on a private road, the rock exposures nearby bear vivid crusts of fresh ocher. The mines left behind a long-lived chemical nuisance: acidic runoff water. To this day, streams and gutters are stained yellow-orange with iron oxides and laced with toxic dissolved metals. The former Leona Heights Mine, at the head of McDonell Avenue, drained acid runoff for more than seventy years before being mitigated. Agents from the Regional Water Quality Control Board monitor these locations, but the water will not soon run clear.

In their heyday the mines attracted geology students earning advanced degrees. At the time, both UC Berkeley and Stanford University, across the Bay, offered mining-related degrees in geology, chemistry and engineering. A generation of academic trainees proved their mettle in the dim-lit workings at Leona Heights. The ore geology of their time was little advanced from the blind gambles of mine operators and the intuition of pick-and-dynamite laborers at the working face. I admire their grit in pushing the frontier of knowledge into the Earth's crust.

Whereas outer space is a vacuum, the inner Earth is the opposite, a plenum. In the vacuum, energy flows at the speed of light and an object set in motion stays in motion. The plenum has no space to move—imagining it feels like being buried alive—and the physics is gnarly, but the plenum is geologically alive, and my tribe is drawn to it. Mines are portals to this realm.

The plenum is beyond direct reach, except through the pinpricks of boreholes and mines. We can mentally follow the tilt of the rocks in a roadcut into the ground. But what we know about the plenum depends mostly on indirect, telluric evidence: how earthquake waves echo, how gravity meters sense buried masses, how electricity circulates in the crust, how volcanic gases smell and groundwater tastes. The deeper we go, the closer we get to the machinery that toys with the Earth's tectonic plates.

I've read some of the papers those graduate students wrote a century ago. They didn't have most of our modern tools, just a young person's energy and the ability to visualize in three dimensions that many geologists have. Each student scientist envisioned something different in the plenum beyond the workings. Given all the digging the miners did from year to year and the disturbances of pyrite fires, each candidate saw a different mine. And each of their theses deserved a degree, but as an ensemble they show me the collective mind in a futile drive against the unknown, decades before plate tectonics somewhat cleared the air.

It didn't help when mistaken ideas were in the way. For instance, some experts held that the pyrite was dissolved from shallow levels and reconstituted just above the water table under so-called vadose conditions. That notion brought a cheeky response from one master's candidate, Henry Mulryan: "If the Leona Orebody is derived under vadose conditions, then it is the only one known to the writer and should take its place in the world's literature on ore deposits." Underground fires kept him almost completely out of the Oakland mines, and he had so little to show for his fieldwork that he threw himself on the mercy of his academic judges with the conclusion, "The writer prefers to await further development in the Leona mine before stating definite ideas."

Science in uncontrolled conditions is hard, and geologists have it worse than most. Mulryan's frustrating struggle is more typical than Schaller's leisurely study of Boehmer's cabinet crystals. Mines have one purpose, and operators can't shut them down for research any more than excavators could halt work to let paleontologists pick fossils inside the Caldecott Tunnel or lay down tools for the soil at the Coliseum site to be sifted for ice-age bones. Science is not all *that* precious.

Much of geologic knowledge is scavenged from places whose owners are getting something else done. And yet, given enough mutually reinforcing threads from enough of these fragmentary findings, geologists can weave a cloth of knowledge that grows stronger with more evidence. Their work moves on and outward from the simple story lines to the frayed margins, where questions push deeper with the aid of more data and better tools.

But not every question can be pursued. The Alma and Leona Mines both encountered thick lenses of nearly pure pyrite, which were consumed to pay for the whole operation. Perhaps a few museums have pieces in a backroom case. More of these marvels surely exist under Leona Heights, but they'll never be found. Foolhardy amateurs were the last people to pluck scattered fragments from the dangerous abandoned tunnels, long before my time.

Early in the rainy season, when the footing is good and the poison oak not yet too lush, I'm sometimes able to pick my way along the deer trails and overgrown tracks in the steep oak woods. On one of those outings, I inspected a small horizontal tunnel—an adit—high in the hills, its existence noted by a mark on a map in a 1910 thesis. The adit was round, about a meter high, and went maybe ten meters into the hillside, untouched since pyrite prospectors dug it out on their knees. I felt a little foolhardy myself. It smelled funny inside, and not even wildlife had found it hospitable— certainly not a mountain lion. Nothing was revealed. Leona Heights is a strange land from a far place and time.

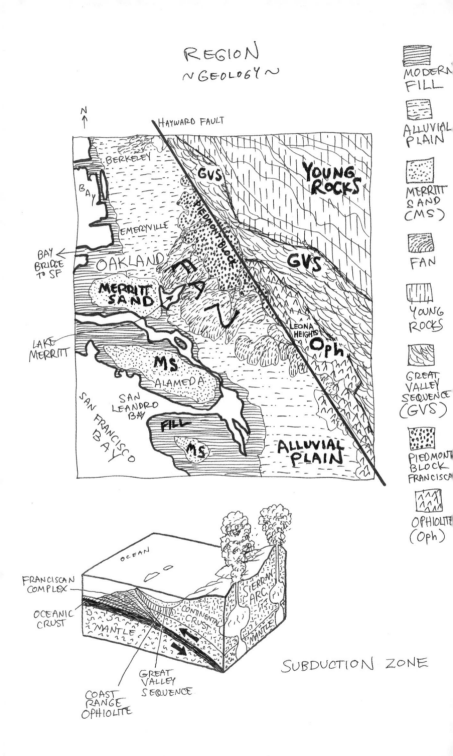

REGION
~GEOLOGY~

MODERN
FILL

ALLUVIAL
PLAIN

MERRITT
SAND
(CMS)

FAN

YOUNG
ROCKS

GREAT
VALLEY
SEQUENCE
(GVS)

PIEDMONT
BLOCK
FRANCISCA

OPHIOLITE
(Oph)

N

HAYWARD FAULT

BERKELEY

BAY

GVS

YOUNG
ROCKS

PIEDMONT BLOCK

EMERYVILLE

OAKLAND

BAY
BRIDGE
TO SF

GVS

MERRITT
SAND

LEONA
HEIGHTS

Oph.

LAKE
MERRITT

MS

ALAMEDA

SAN
LEANDRO
BAY

SAN FRANCISCO BAY

FILL

MS

ALLUVIAL
PLAIN

OCEAN

FRANCISCAN
COMPLEX

OCEANIC
CRUST

MANTLE

CONTINENTAL
CRUST

SIERRAN
ARC

MANTLE

COAST
RANGE
OPHIOLITE

GREAT
VALLEY
SEQUENCE

SUBDUCTION ZONE

11

THE RIDGELINE

Oakland was named for a live-oak forest by the Bay, but today the city's defining feature is the high hills. The Oakland Hills are young landforms, only a million or two years old. They owe their existence to the Hayward Fault, thanks to the transpression that steadily raises the eastern side about a millimeter per year. As the rising hills erode, they expose an exceptional range of rock types that have been mapped into some dozen named formations. I introduced the youngest set at Sibley and the oldest in Leona Heights. Time for the stories inside the rest of the hills, and the larger tale in which they all fit.

Oakland's high landscape, it should be plain by now, is strongly tectonic. Many different slices of rock have been shuffled and dealt to assemble this unusually complex range of hills. They don't offer us a stately parade of level strata like the Grand Canyon, or marching hogbacks of tilted strata like the ridges of Shenandoah. They aren't the craggy Grand Tetons or New England's loose gatherings of noble, time-worn peaks. Instead, their strongly layered rocks have been squashed and stretched lengthwise, and turned nearly vertical, parallel to the ridgeline. Their structure gives the hills a peculiar and distinctive look: Oakland beautiful.

Each unit of rocks in the hills is a chip in a great crustal mosaic with patterns and themes that ramify through California, the western United States and much of North America—all in our backyard. And there's much we might yet learn from them.

As it happens, the highest ridgeline runs at a slight diagonal through these rocks, such that the ridgetop roads, Skyline and Grizzly Peak Boulevards, touch nearly all of the mapped formations. These ridge roads along the eastern city limits were built in the early 1900s to attract residents. The newspapers praised them as pleasure drives, great places to take that first family automobile. Today the roads, and the parks along them, give good access to the high hills.

Seen from the city below, especially from the Bay shore, the hills may seem like an even wall, or a wave of rock carpeted with woods. But look again when the air thickens with marine haze and they resolve into a series of en echelon ranges, a set of waves with profiles as crisp as if cut from paper. They have landmarks seen all over town: Grizzly Peak, Round Top, Redwood Peak, Sugarloaf Hill, Fairmont Ridge. In front of the main ridge are the hills of the Piedmont block and the looming breaker of Leona Heights; to its rear are some thirty miles of ridges of similar height that together make up the Coast Range.

Also visible from below are knobs and knees that project toward the Bay, some of them draped with streets and some with woods. Between these promontories are valleys, some large enough to be called canyons. Each valley hosts a stream, and each stream is fed by water from the hills, either rainfall wrung from uplifted sea air or dew sifted by trees from the summer fog. The hills collect the water that cuts the valleys and nourishes the flats.

The hills are an arena of competing Earth energies,

tectonic forces raising the heights and erosional forces cutting them down. They present a landscape always under construction, a dynamic equilibrium that comes of perpetual interplay among agents of geologic action. To the geologist, mountain ranges are shape-shifting places of rapid and continual decay. Between stream erosion and landslides, the Oakland Hills are falling down. That wasn't a problem for the city until thousands of families moved into them.

The hills form a topographic barrier with political significance; they define the city and county boundaries and rebuff the builders of roads. The main ridge, with its rises and saddles, is consistently higher than a thousand feet. Beyond it to the east is a wildland belt reserved for parks and watershed, followed by the wealthy exurb towns of Orinda and Moraga in Contra Costa County. The backbone ridge was historically a pernicious barrier that only three tunnels, two of them long ago closed and sealed, have penetrated and just four small roads have crossed. Today the Caldecott Tunnel, north of the Sibley preserve, is the only way through, not over, the hills. But that misses most of the geology.

People who see the hills in old photographs often remark on how bare they looked. Even a century ago, it was a common misconception that the hills were denuded by logging. But for thousands of years they were treeless meadows, visited seasonally under Indigenous management, and the old photos show remnants that survived the Spanish and Mexican occupations.

The one exception was the redwood groves of the southern Oakland Hills, a restricted forest that extended a few miles eastward from upper Dimond Canyon over the ridgetop to the outskirts of Moraga. Some of the trees, to judge from their colossal stumps, were among the largest of their kind, rivaling the giant sequoias of the Sierra Nevada.

In the early 1800s, a cluster of redwoods on the highest ridgeline "too conspicuous to be ignored" was a navigational aid for sailors threading the mouth of the Bay; aligning the trees with the northern tip of Yerba Buena Island assured safe passage past notorious Blossom Rock.

The tribes groomed the redwood stands with their customary fire but didn't need the timber. The Spanish protected them as crown property, and only small trees were taken to build the Mission San José in the 1790s. But after sovereignty changed from Spain to Mexico, foreign infiltrators staged a redwood rush in the 1840s that lasted about fifteen years, mowing down the ancient trees with characteristic abandon. When the trees were gone, gleaners mined their stumps for firewood and shingles. But redwoods are tougher than lumberjacks, and a second-growth forest has come into its own almost two centuries later.

In Oakland's early days, the hills were remote ranchland served by a few wagon roads. Then they became the last frontier of land speculation for private water companies and syndicates of real-estate men. Some landowners started and abandoned a timber scheme that involved planting millions of blue gum eucalyptus trees, an invasive and fire-prone species that still plagues the heights. The eccentric celebrity writer Joaquin Miller planted some seventy-five thousand trees on his hillside acreage, now part of the city's Joaquin Miller Park. Then developers and residents planted thousands more as the hills filled with homes.

By 1920, the developer's template had changed from streetcar suburbs to automobile suburbs. A middle-class white family could own a car and a house, living high above it all in a modest version of a Piedmont grandee's estate. This template, a fantasy that endures as "the American dream," pushed a wave of suburban sprawl to the crest of the Oakland Hills. For the first time, they glowed with lights at night and reflected in Lake Merritt along with the stars. And with that driving fantasy the Realty Syndicate's dream came true: there was gold in them, their hills.

The residential invasion stopped at the ridgetop, on the edge of the watershed lands behind it. That property, the most valuable asset of the floundering private water companies, was gathered under the ownership of East Bay MUD in 1928. The water agency sold parts of the land to the newly formed East Bay Regional Park District, which established today's great belt of ridgetop parks.

As developers advanced into the hills, the ground began to matter. In the geologist's speeded-up movie, the hills are melting and spilling and sagging down as fast as they're being tectonically lifted. The hills want to slough us off, and landslides are a constant threat.

To their victims, landslides feel personal and loom large. To the geologist, landslides are surficial chisel chips that shape the heights. On the million-year time scale, they sweep away every piece of hillside. They generally leave land unfit for human use; their scars, visible on the map as vacant lots and phantom rights of way, may take lifetimes to heal. Yet there are deep-seated landslides in Berkeley so large and gentle that people live on them, even as their homes drift over the property lines.

Geologists refer to the "critical zone," the realm between vegetation and virgin bedrock, where soil, groundwater, life and slowly decaying rock coexist in a delicate balance of saturation, pore pressure and gravity. On steep hillsides, the critical zone undergoes soil creep, a quiet slumping process that tilts trees and warps fences. Prolonged rainfall nudges it all toward failure, and earthquakes too can give it a final push.

As trees have spread over the former grassland, the critical zone in the Oakland Hills has become drier on the whole, and thus a bit more stable. Trees draw water out with their roots, and they also catch rainfall on their leaves and bark, where the intercepted water evaporates before it can reach the ground. But at the same time, the high car suburbs have raised the ambient risk as roads and foundations cut into the critical zone, diverting runoff into the ground and undermining the slopes. The hillsides have responded with a rash of landslides. And those are the ordinary kind, triggered by wet winters. Earthquake shaking can launch thousands more at any time of year.

Any location in the Oakland Hills, then, can expect to be destroyed, sooner or maybe centuries later, by landslides and earthquakes. Buildings that escape them have merely beaten the odds for now; cities that regard them lightly, as Oakland did a century ago when the suburbs were permitted and built, are letting people put their lives in fate's way.

• • •

On the positive side for geologists, the same forces that tear down the hillsides reveal the bedrock. Outcrops hide among the trees, and roadcuts offer lots of exposures to study and enjoy. The rocks exposed in the hills range in age from about 10 to about 160 million years, from the late Miocene at the Sibley preserve to the late Jurassic at Leona Heights.

In the truly big context, the different rocks of the Oakland Hills all represent aspects of plate tectonics. The billion-year history of the plates looks like a series of breakups and mergers with only hints of rhyme and rhythm. But from about 175 to about 25 million years ago, from the middle Jurassic period to the early Miocene period, California was a long-lived subduction zone. All the rocks on the south end of the ridgetop date from that time.

Subduction, the tectonic engine driving the Earth machine, makes room for the plates to move around. It varies in style, but the geometry is the same: one plate sinks beneath another plate. The subducting plate is the thin kind, oceanic lithosphere, made of dense black lava and often covered with a thick coat of seafloor mud. The overriding plate may also be oceanic: think of the long Aleutian Island chain, where the Pacific plate dives under oceanic lithosphere in the Bering Sea. Or it may be continental, a thick body of less-dense granite: think of mountainous Chile, where South America overrides the oceanic Nazca plate. In either case, because oceanic plates are involved, the action takes place beneath the sea.

The sinking plate rubs its top against the base of the overriding plate in a broad, gently sloping fault zone called a megathrust. With their large surface areas, megathrusts build friction and release it in the Earth's greatest earthquakes. The megathrust may even shave slices off the sinking plate.

As the sinking plate heats up and feels growing pressure, the seawater soaked into it is squeezed out, and next the water that chemically bonded with its minerals is expelled. This hot, mineral-charged juice rises into the upper plate and induces the rocks to melt into magma. Whole blobs of seafloor sediment can rise up too, like underground balloons, and either lodge in the upper plate or turn to magma.

The magma, in turn, rises and erupts in curving lines of volcanoes. The roots of these volcanic arcs melt and freeze repeatedly as pulses of magma rise through them. As the deep magma slowly cools, it forms complex bodies of granite that may become exposed much later as the volcanoes erode away. Erosion crumbles the mountains into sediment, which washes down to sea and is deposited on the seafloor.

Subduction, magmagenesis, volcanism, erosion, deposition—these form the cycle in what geologists call the subduction factory. The rocky seafloor crust sweeps back into the mantle, and its volatile topping rises back to the surface. Water mostly returns to the surface, but some goes down with subducting plates to re-fertilize the mantle. Continents are stirred and rejuvenated. The machine keeps the world young.

Subduction is hard to study because it always occurs under the ocean. But the California subduction zone got interrupted, broken and lifted above the sea in pieces we can partly reassemble. It has attracted researchers for over fifty years.

The south end of the ridgeline runs through rocks of the California subduction zone for about seven miles. The ages of these Oakland rocks span roughly fifty million years, with wide gaps in that stretch, but the subduction zone persisted for well over a hundred million. That's a long time, rare luck, even to a geologist.

The textbook subduction zone taught in schools has a deep-sea trench on the ocean side where the plates first meet, a line of volcanoes on the land side where the magma comes up, and between them a wide zone of ocean floor—a foreland basin, as geologists call it—that collects sediment from the volcanoes and the land. Each of these parts makes characteristic rocks.

The California subduction zone varied over its lifetime, but it was usually a belt about a hundred miles wide between the North America plate on the east and the incoming oceanic plate, or plates, on the west. The western edge, along the trench, produced a scramble of rocks called the Franciscan Complex. The ophiolite lay next to it; as I said, there are several different stories about just where it was and how it got there. The basin in the middle collected a thick stack of rock beds called the Great Valley Sequence; most of it still fills the Central Valley. On the east, the volcanic range came and went several times; its roots, exposed by erosion, are the white granite of the Sierra Nevada. The ancient coastline lay far east of today's, at the position of the Sierra foothills.

Pieces of all four parts of the old subduction zone are present in Oakland. The Sierran granite, if I may cheat a little, is in City Hall. Franciscan rocks underlie everything west of the Hayward Fault, including the uplifted Piedmont block; the ophiolite crops out at Leona Heights and elsewhere in the western half of the southern Oakland Hills; and the Great Valley Sequence makes up the eastern half, including the ridgetop.

Someone had to map these rocks before their stories could be extracted. Geologic maps are more than psychedelic stripes and slabs of color dotted with peculiar symbols; they're visual records of intensive work by specific people whose presence I feel as I stand at the rocks represented on the maps. We are lucky to have a century's worth of investigation summed up in bedrock maps of Oakland made freely available by the US Geological Survey, most recently in 2000.

The southern ridgeline contains three different slices of the Great Valley Sequence, separated by faults, that mappers have divided into a half-dozen formations. The rocks vary from smooth shale to rugged conglomerate, their colors a limited range from dark greenish brown to light golden brown. We know their ages fairly well, but not exactly where they were before tectonics brought them here. They might have originated as far south as present-day Mexico. None have visible fossils except the oldest: the Knoxville Formation, centered around Knowland Park in the southernmost hills, has a few late Jurassic versions of clams and squid in its dark mudstone.

The most noteworthy of these formations, the Oakland Conglomerate, holds up the ridgeline for its southernmost five miles and continues south another thirty. It's full of cobbles, the size and shape of potatoes, in an even matrix of fine Sierran sand. The cobbles consist of a wide variety of rocks, the record of a former Sierra unlike our Sierra.

Along Skyline Boulevard, the Great Valley Sequence is interrupted by about a mile of blue-green serpentine rock belonging to the ophiolite. The change is dramatic as the trees disappear and wide views open to the east and west.

Earlier I called Franciscan melange California's most distinctive lithology, but serpentine is a close contender and is officially the state rock. It began existence as the mantle rock at the very base of the oceanic plate, but infiltrating superheated seawater hydrated its dark crystals into glossy sea-green minerals. Missing key nutrients and often high in toxic metals, serpentine makes a thin reddish soil that repels most ordinary plants and supports many rare and peculiar species that can take the conditions. Serpentine Prairie, part of Redwood Regional Park, preserves the back side of the ridge; on the Bay side the Crestmont gridiron of 1950s ranch-style homes exposes the serpentine in its high roadcuts. More of this rock runs through Joaquin Miller Park, where it stunts the trees Miller planted in it.

North of the serpentine are two more miles of Great Valley Sequence rocks, including the two younger slices. Most notable in these is the Redwood Canyon Formation, a band of sandstone that underlies most of Oakland's redwood groves as well as Redwood Peak. The strong but porous bedrock, combined with the area's elevation and local weather, offers a footing moist and stable enough for redwoods to live thousands of years. It also yielded the stone used in the First Unitarian Church in downtown Oakland. The uppermost, northernmost slice is a nameless mudstone of Eocene age, perhaps about forty million years old. It's too young to be officially in the Great Valley Sequence, but it too was part of the basin known today as the Central Valley.

The whole time these rocks were being laid down, the subduction zone was in full swing. It's important to note that because the Earth was very warm, without glaciers and ice caps, the ocean stood hundreds of meters higher than today. A large oceanic plate that geologists have named the Farallon plate was converging against the North America plate, from southern Canada to halfway down Mexico. It wasn't always a calm, steady process. At times when the Farallon plate dropped steeply down into the mantle, the Sierra Nevada volcanic arc would have a session of vigorous activity, recorded today in its deep granite roots and the sediments it shed to the west.

Sometimes instead the Farallon plate descended at a shallow angle in what's called flat-slab subduction, getting in the way of North America as the two plates converged. Around ninety million years ago a fat volcanic plateau rode into the trench and pushed around the overlying crust like a puppy burrowing under a throw rug. The Sierran arc shut down and the crust behind it rumpled up in a high plateau, and whole sheets of rocks slid off the plateau to the east, raising more mountain ranges in what geologists call fold-and-thrust belts.

At that time, this part of western North America

looked much like western South America today, like Peru and Bolivia's Altiplano plateau and the fold belt to its east. Geologists have named North America's version the Nevadaplano. As the Farallon plate moved on and sank, the Nevadaplano slowly relaxed and stretched apart. Volcanoes came forth in waves and showered the broken plateau with blankets of ash.

Most of Oakland's rocks of the Great Valley Sequence record outpourings of sediment from the ancestral Sierra Nevada and neighboring regions. The younger Eocene mudstone may also include sand carried here from Idaho by a long-lost river. At least, that's what rocks of the same age farther east tell us. But everything I say about Oakland's rocks is tentative because their original positions are poorly constrained.

Before I go on, I want to revisit the Franciscan Complex, which finally makes sense as part of the California subduction zone. Like other subduction complexes, it formed by accretion, a general-purpose word for adding new material onto old material. Specifically, it accreted mostly by underplating. As the Farallon plate approached the trench, the subduction fault kept paring off slices of it that stuck onto the underside of the upper plate, putting younger rocks *beneath* older ones, the opposite of how Nicolaus Steno said rocks are made. Picture the machine in an old movie sliding freshly printed newspapers onto the back end of a growing stack—only the subduction zone mixed neat slices of oceanic lithosphere and mangled chunks of stuff.

Large landslides delivered random piles of Franciscan rocks into the trench—old and young rocks, raw mudstone and high-grade blocks—and that's how the subduction zone made melange. This lithological scrapple confused California geologists for a hundred years, until the advent of plate tectonics in the late 1960s.

• • •

Oakland's younger set of rocks, on the northern part of the ridgeline, formed after the subduction zone broke up. The breakup started about twenty-eight million years ago, when the irregular spreading ridge at the back end of the incoming Farallon plate arrived at the trench, which ran out of stuff to subduct. Instead North America found itself with its nose against the far side of the spreading ridge, the Pacific plate, which was being pulled northwest. When that happened, the relative motion between the plates abruptly shifted from convergent to transcurrent, and that's where and how the San Andreas Fault system formed, tugging the western edge of California sideways. The old California was dismantled as today's California took form, in Miocene time.

The change in motion wrinkled up the Coast Range as the pieces of the former subduction zone were rearranged. Farther east, the breakup of the Nevadaplano intensified, and major episodes of explosive volcanism left ash deposits from distant places within the greater Bay Area's younger rocks. As the Earth's crust adjusted in the mountain states, the modern Sierra Nevada range rose and shut off the moisture supply to the Great Basin, to California's advantage.

Rocks from this phase of Oakland's geologic history are exposed along the ridgetop for eight miles to its northernmost end on Oakland land, near Grizzly Peak high above Berkeley. The rocks grow steadily younger in a northward trend. The oldest part is a fragile, dark-brown mudstone traditionally assigned to the Sobrante Formation. At roadcuts it steadily buries roadside signs and hydrants in tiny clay-rich fragments, and trees above the crumbling slopes threaten to fall. The ridgetop dips in a pass as it intersects this rock, where Shepherd Canyon and Pinehurst Roads meet. This rock formed near land in a seafloor basin roiling with fresh fine-grained mud and clay.

Three miles of tall roadside bluffs expose the next-younger unit, the steeply tilted Claremont Shale. This formation generally consists of shale, hence the name, but in Oakland it's pale, flinty chert in layers a few inches thick alternating with thinner layers of soft brown shale. The layers, thousands of them, stand on end like a storm-swept forest of golden bamboo. This belt of rock underlies the Huckleberry Botanic Regional Preserve, where its scant nutrients and sharp drainage create some rare habitats. It also underlies some of Oakland's highest residential streets and scenic roadside stops.

Chert like this forms in sheltered ocean basins in nutrient-rich water, where the microscopic bodies of diatoms can rain down on the seafloor undiluted by mud from the land. Diatoms are one-celled plants with shells of opal that make a drop of oil inside their shells to help them float. In thick seafloor deposits of diatom ooze, the opal shells merge to form a hard, opaque stone with a surface like wax or porcelain, and their oil drops drift free. The Claremont Shale is rock of the same kind and vintage as the enormous Monterey Formation, source of much of California's oil and gas. Underground the rock is black with hydrocarbons, but in most outcrops the oil has oxidized away and left it bleached blond.

The Claremont Shale's rhythmic bedding may result from a self-organizing process as the diatoms changed from seafloor ooze to layered rock. Such a process may also have created the occasional pods of dolomite found in this rock unit.

The upper and lower boundaries of the mudstone and chert are not transitions but dislocations. They are faults.

These bodies of rock have been shuffled. The sequence they make is not a story, more like a few spilled storyboards from a movie treatment. They tell of several different parts of Miocene California that were much like today's Southern California, a confused mix of islands and ranges and faults.

The chert gives way to the younger rocks I described at the Sibley preserve, starting with a dip at the head of Claremont Canyon, where the telegraph line once ran over Summit Pass. The conglomerate of the Orinda Formation, less resistant than its neighbors, underlies this saddle. Last to appear are the hard lava beds and ash layers of the Moraga Formation, forming highlands and overlooks that the locals visit to watch the sun set past the Golden Gate.

Our high hills expose snapshots from all of Oakland's geohistorical stories, from the subduction zone's rock factory to the later transpression that broke, chewed and smeared it all sideways—topped with real live volcanoes.

We owe these stories to generations of geologists who recorded their findings in papers and maps. The maps are especially precious to me because the hills, like Mountain View Cemetery's gravestones, are perishable even in human time. Consider a small body of rock a few hundred meters across halfway up Snake Road. It was pure luck that the geologist who mapped it could assign these rocks a Paleocene age, roughly sixty million years old, on this basis: coral fossils in a foundation hole reminded his field partner of Paleocene corals in a Santa Cruz Mountains roadcut. Today, a quarter-century later, that roadside hole is decayed and overgrown. I've come and inspected the spot closely several times, and whatever they saw is no longer visible. Someone needs to keep an eye open for fresh exposures as roadcuts are renewed and new foundations dug.

The Earth moves on, and not just in quiet ways like

this. New maps will need making when large landslides change the hills, when the rising sea alters the shoreline and when the Hayward Fault reveals its active trace in the next major earthquake. Mapping also never ends because knowledge grows and the minds of geologists advance with their science. They have new questions to ask the rocks. The leading edge of science is unmapped, a collective stew of thought in the minds of scientists, shared with colleagues in face-to-face conversations and papers in journals. I keep up with them as best I can because I love their quest.

Oakland is just one place in a wide world, but one exceptionally rich in evidence of deep Earth history, deep Earth processes and deep human changes. With increasing study, our rocks and landscape have revealed geological stories extending ever wider across the state and continent, just as the city's history is a story of growing political stature, commercial influence and environmental impact. This complex city has grown on complex ground.

From the Bay to these hilltops, the Ohlones used every tool they knew to shape this landscape, which once constituted their world. Today they share their land, now greatly changed, with the rest of us in a system of civilization that not only encompasses the whole planet but increasingly perturbs it.

Geology gives us tools to understand the Earth in new depth, and our lifeways, which depend so much on the Earth and its resources, need to take into account the knowledge those tools have brought us. What we do in the generations to come, with the seventh generation in mind, must respect not just what the plants and the animals have told the Ohlones but what the Earth, with growing urgency, is telling us. Oakland is a good place to start listening.

ACKNOWLEDGMENTS

To make this book I have walked the land of Oakland on my own, but guided by the wisdom of generations of Earth scientists. Here I'll mention only those I've met face to face. Some are now gone; others have long careers ahead of them.

Earth scientists of three great Bay Area institutions have helped me understand Oakland better, whether they know it or not. From the US Geological Survey I thank Brian Atwater, Earl Brabb, Bill Brosgé, Terry Bruns, Paul Carlson, Mike Diggles, William Glen, Russ Graymer, Jon Hagstrum, King Huber, Rudy Kopf, Vicky Langenheim, Virginia Langenheim, Jim Lienkaemper, Julian Lozos, Bob McLaughlin, Bruce Molnia, Walter Mooney, Bonnie Murchey, Tor Nilsen, Michael Rymer, Phil Stoffer, Nancy Tamamian and Ray Wells. From Stanford University I thank Kat Burnham, Allan Cox, Gary Ernst, Norm Sleep, Nick Van Buer, and Mark and Mary Lou Zoback. From UC Berkeley I thank Walter Alvarez, Maria Brumm, Raymond Jeanloz, Michael Manga, Doris Sloan, Nick Swanson-Hysell and Clyde Wahrhaftig. All three institutions support excellent libraries as well.

Other geoscientists I cannot overlook, because they did not overlook me, include Don Anderson, Tanya Atwater, Joyce Blueford, Wallace Bothner, Sam Carey, Steve Edwards, Rolfe Erickson, Gillian Foulger, David

Grinspoon, Warren Hamilton, Brian Hausback, Susan Hough, Lucy Jones, John Karachewski, Tom MacKinnon, Eldridge Moores, Dave Mustart, Jason Saleeby, Anne Sanquini, Cecil Schneer, Ray Sullivan, Jeff Unruh, John Wakabayashi and Terry Wright—especially Terry for the breadth of his pleasure and the depth of his commitment to sharing it.

My two local geological societies, the Northern California Geological Society and the Peninsula Geological Society, have enriched me with their meetings and field trips, and my two national societies, the American Geophysical Union and Geological Society of America, have also done so with their great annual meetings and their invaluable journals and books.

I have been stimulated, challenged and encouraged by the geoscientists, historians, writers, enthusiasts, polymaths and citizens of Twitter, Facebook and especially the Well. The online world has combined "reading, conference and writing" just as Francis Bacon aphorized in the seventeenth century and Vannevar Bush envisioned in the twentieth. Both of those great utopians would have celebrated today's internet. Let me single out the David Rumsey Historical Map Collection and the Internet Archive for special praise and gratitude during this time of pandemic when access to so many resources was restricted.

From Oakland itself I must first thank Dennis Evanosky, whose thorough review clarified major and minor historical matters. I'm also grateful for the deep resources made available by Amelia Sue Marshall, the Oakland Heritage Alliance, the Oakland and San Francisco public libraries and the Friends of Sausal Creek.

Portions of this book were reviewed by Gene Anderson, Tom Barry, Greg Bartow, Michael Colbruno, Cian

Dawson, Sandy Figuers, Morgan Fletcher, Paul Henshaw, Dorothy Londagin, Tom MacKinnon, Amelia Sue Marshall, Don Medwedeff, Katie Noonan, Liam O'Donoghue, Naomi Schiff, Will Schweller, Mark Sorensen and John Wakabayashi. I thank them all.

For help on the way, I gratefully acknowledge Oakland's brewers and cannabis suppliers, its public transit operators and its generations of journalists. Finally I thank this city's engaged citizen organizations, whose members have let me speak to them, wave my arms and show them around.

Every writer benefits from companionship and advice, and for those I thank the science editors of KQED who supported me, the journalists of the Northern California Science Writers Association who inspire me, and the growing community of scientists who share their work and their worldview with the public. Scott Kurnit provided my entree, after many years of simmering, to the writing life. And I am greatly indebted to Marthine Satris of Heyday for her gentle, skilled and insightful guidance, art director Diane Lee and copyeditor Michele Jones, as this manuscript took shape.

I could have not found a better artist than Laura Cunningham to convey the blend of scientific insight and real-world accessibility this book demands. Her contribution was immeasurable and our collaboration most pleasurable.

Lastly I thank some extraordinary people: my parents and their parents and my siblings, who made me what I am, and my wife, Fleur Helsingor, who anchors me.

NOTES

In this book I summarize a great deal of scientific information based on several decades of immersion in the geologic literature. I cite specific sources for advanced material, leaving more basic topics to the usual reputable websites. I also cite illustrated posts from my *Oakland Geology* blog (https://oaklandgeology.com/). The geologic map of Oakland there (https://oaklandgeology .com/oakland-geologic-map/) is derived from Graymer (2000). All quotes from newspapers have been verified on https://www .newspapers.com.

1 THE HAYWARD FAULT

Lienkaemper (2006) is my authority for the active trace of the fault, and McFarland, Lienkaemper, and Caskey (2016) provide current monitoring data for the fault. Louderback (1947) collates historical reports on earthquakes of the 1830s, and Boatwright and Bundock (2008) compile a thorough set of reports from the 1868 earthquake. Hough and Bilham (2006) explore the wider ground of historic earthquakes, their effects on cities and their aftermath.

Stonewall Road See the scene at https://oaklandgeology .com/2008/06/19/the-fault-at-stonewall/.

I picture the next major earthquake The US Geological Survey's HayWired project (https://www.usgs.gov/centers/wgsc /science/haywired) describes a future major earthquake on this fault in scientifically sound detail.

"successions of dairies and of reservoirs" Jordan (1907), 5.

shutter ridges See this landform at https://oakland geology.com/2008/07/31/shutter-ridges/.

En echelon cracks See some at https://oaklandgeology .com/2007/09/29/the-oakland-fault/.

Records from this creepmeter See running records for station CTM at https://earthquake.usgs.gov/monitoring /deformation/data/plots.php and see physical details of the site at http://ciresi.colorado.edu/~bilham/CREEPDATA/Temescal %20Park.htm.

the south side of 39th Avenue See the site at https://oaklandgeology.com/2011/05/27/fault-gauge-39th-avenue/.

Researchers come here on a schedule See their data at http://funnel.sfsu.edu/creep/.

movements of the seven giants who hold up the world Guinn (1907), 51, quoting a letter from Hugo Reid.

earthquake of 8 December 1812 Known as the Wrightwood or San Capistrano earthquake, it ruptured the southern San Andreas Fault.

scientists uncovered its latent trace there Lienkaemper and Williams (1999) is the report of their trenching study.

The Hayward Fault shrugged Lawson (1908, vol. 2) compiled reports of the 1868 earthquake four decades later. Other sources include Halley (1876) and contemporary newspapers.

"rolled with huge waves like the sea" Lawson (1908), 2:442.

"the backs of a drove of buffalo" *Napa Reporter*, quoted in *San Francisco Chronicle*, 26 October 1868.

"the mountains seemed to skip like" *San Francisco Chronicle*, 22 October 1868.

"surge to and fro and champ at their hawsers" *Sacramento Bee*, 21 October 1868.

"trees whipt about like straws" *Daily Alta California*, 22 October 1868.

"rushing down the bed of the creek" *Alameda Democrat*, 24 October 1868.

"a young man of much promise and ability" *Daily Alta California*, 22 October 1868. Halley (1876) recounted his funeral.

"it is well braced with iron" *Oakland Daily Transcript*, 22 October 1868.

"desultory conversational character" Rowlandson (1869), 4.

A rumor persisted The rumor is documented by Prescott (1982) and refuted by Aldrich et al. (1986).

"after a careful analysis" *San Francisco Chronicle*, 22 October 1868.

ripped apart the adobe home of Don Guillermo

Castro *Hayward Review*, 20 May 1930.

    a change from eyesore to amenity Toké, Boone, and Arrowsmith (2014) assess the added value of fault-related greenspace.

    "usually I do not think rapidly" Mark Twain, "The Great Earthquake in San Francisco," *New York Weekly Review*, 25 November 1865.

    a wavy surface like that of a theater curtain Phelps et al. (2008) present a three-dimensional model of the fault.

    the three faults are effectively one Harris et al. (2021) provide simulations of three-fault ruptures.

    the last dozen large quakes on the Hayward Fault Lienkaemper, Williams, and Guilderson (2010) document a nineteen-hundred-year history of the fault.

    Revere Avenue See https://oaklandgeology.com/2016/07/11/marks-of-the-oakland-fault/.

2 LAKE MERRITT

The Lake Merritt Advocates website has a detailed timeline of the lake's human history at https://www.lakemerritt.org/history.html. Gary Lenhart has a gallery of historical postcard images at https://alamedainfo.com/lake-merritt-oakland/.

    The earliest good map of this area See and download the 1857 Bache map at https://en.wikipedia.org/wiki/Oakland,_California.

    These two platforms, so much alike See them on the geologic map at https://oaklandgeology.com/2016/08/22/the-merritt-sand/.

    One might start with the sediment See the gravel at https://oaklandgeology.com/2008/07/15/adams-point/.

    approximately 125,000 years Helley, Fitzpatrick, and Bischoff (1993), 3.

    "to prevent the destruction of fish and game" See the law's text at https://www.lakemerritt.org/150th.html.

    the growing town's untreated sewage fouled the water Baker (1914), 146–49.

    an early settler named Romby dug clay there *Morning Call* (San Francisco), 19 September 1893.

the display of suiseki See it at https://bonsailakemerritt
.com/garden/suiseki-display/.

"First there is a mountain" See the origin of this lyric by
the songwriter Donovan Leitch at https://en.wikipedia.org/wiki
/There_Is_a_Mountain.

**the Pacific coastline lay far out by the Farallon
Islands** See a map and animation at https://animations.geol
.ucsb.edu/2_infopgs/IP2IceAge/cSFBayFlood.html.

it will be a meter, maybe two, higher than today See a
visualization at https://oaklandgeology.com/2018/01/08
/walk-around-lake-merritt-after-sea-level-rise/.

3 DOWNTOWN

Bagwell (2012) is the standard introduction to Oakland's civic
history; Halley (1876), Wood and Munro-Fraser (1883), Guinn
(1907), Baker (1914), Merritt (1928) and Conmy (1961) are valuable sourcebooks. Promotional books including Scott (1871), Elliott (1885), Oakland Merchants Exchange and Oakland Board of
Trade (1897) and Blake (1911) document the flavors of Oakland
boosterism prevailing at different times. Schwarzer's (2021) critical modern treatment of Oakland's development since the 1890s
is essential reading.

"the only objects wanting" Beechey (1831), 2:4.

three men in 1850, armed with gold Halley (1876),
447–53, offers a caustic summary of Oakland's founding.

"from its occurrence on Lake Merritt" Lawson
(1914), 15.

three separate lobes See these and their offshore exten-
sions at https://oaklandgeology.com/2020/02/03/the-merritt
-sand-a-little-deeper/.

"acute and obtuse angles" Wood and Munro-Fraser
(1883), 563.

"unscrupulous grabber" Halley (1876), 203.

"the chief ornament and attraction of this city" Wood
and Munro-Fraser (1883), 562.

"what was once a fine grove" Cronise (1868), 150.

"gnarled veteran" *Oakland Tribune*, 17 January 1917.

"Marshall Curtis raised carrots" Oakland Merchants Exchange and Oakland Board of Trade (1897), 27.

now a museum See https://pardeehome.org.

"monument to metropolitan progress and energy" *Oakland Tribune*, 26 June 1910.

"a celebrated California building material" *Oakland Tribune*, 26 June 1910.

The city preferred it over sandstone *Oakland Tribune*, 3 June 1911.

City Hall is a trend-setting case study Walters (2003) reviews this history-making retrofit.

"verd antique" See more about the quarry at https://www.vtverde.com.

dolomite rock from the Natividad Quarry See more at https://oaklandgeology.com/2016/07/18/oakland-building -stones-kaiser-centers-dolomite/.

local sandstone, from the small McAdam quarry See more at https://oaklandgeology.com/2022/04/11/mcadams -quarry/.

"quarrymen were unable to deliver this material in sufficient quantities" Wendte (1927), 90.

4 MOUNTAIN VIEW CEMETERY

The cemetery's website is at http://mountainviewcemetery.org. Evanosky (2007) details the cemetery's founding and growth and also introduces its styles and personages. Michael Colbruno's *Mountain View People* blog (https://mountainviewpeople.blog spot.com/) features biographies of hundreds of people buried there.

"there are certain laws underlying all development" Le Conte (1879), 266.

"gruff at times" See Taliaferro's biography at https://eps .berkeley.edu/content/nicholas-lloyd-taliaferro.

He would bring classes to the hills See another version of this oft-told story by Erik Vance in "School of Rock," *Bay Nature*, 2010: https://baynature.org/article/school-of-rock/.

And here are the native rocks See their portraits at https://oaklandgeology.com/category/cemetery-boulders/.

In founding the Sierra Club See documents from the Sierra Club's early years at https://www.sierraclub.org/library /history-archives, under revision at the time of this writing.

Le Conte first visited the Sierra in 1870 See Le Conte's edited diary at http://www.yosemite.ca.us/library/leconte/.

"beautiful are the things we perceive" Hansen (2009) offers a discussion of this saying.

"a way must be opened for the human understanding" Jones (1937), 242.

"the pursuit of science as a private business is a losing game" Cooper is quoted at https://en.wikipedia.org/wiki /James_Graham_Cooper.

5 THE PIEDMONT BLOCK

The Piedmont Historical Society (https://piedmonthistorical.org) hosts an annotated pictorial history of the town, and a recently established website by a Piedmont resident, *History of Piedmont, California* (https://www.historyofpiedmont.com/) is an ambitious, independent compilation of historical materials. The Franciscan Complex is not easily summarized; Raymond (2017) takes sixty-two pages, Wakabayashi (2015) takes seventy-seven, and even their abstracts are opaque to lay readers.

"Sheared black argillite" Graymer (2000).

a bold outcrop of dark, waxy, wavy-banded chert See it at the end of https://oaklandgeology.com/2020/10/26 /upper-indian-gulch/.

Oakland's old macadam rock was probably recycled See some at Lake Merritt at https://oaklandgeology .com/2014/12/23/our-local-fill/.

The old main pit, partly filled in See the former Blair Quarries at https://oaklandgeology.com/2021/12/06/blair -quarries-not-the-same-as-blairs-quarry/.

Two other reservoirs, of the same type and vintage See these at https://oaklandgeology.com/2019/08/19/oakland -geology-ramble-8-piedmont-ridge/.

"considerable local reputation as medicinal waters" California State Mining Bureau (1894), 331.

ten times the average levels in today's tap water See

annual water quality reports for Oakland at https://www.ebmud
.com/about-us/publications/.

"The water of both springs is noticeably sulphureted"
Waring (1915), 269.

small and fleeting signs of unusual mineral content
See more at https://oaklandgeology.com/2013/09/08
/piedmont-sulfur-spring/.

**"the cars on the tracks are standing as they were
twenty years ago"** *Oakland Tribune*, 17 December 1890.

For now, it's a dramatic place See more at https://oak-
landgeology.com/2011/08/10/hayes-creek-dracena-park-walk-34/.

This was recognized a century ago See Buwalda (1929).

an unusual subvolcanic lava See more about this quartz
diorite at https://oaklandgeology.com/2020/07/06/rocks-of-the
-bilger-quarry/.

"on a windy day in summer" *Oakland Daily Transcript*,
23 March 1873.

these two outfits were a duopoly *Oakland Tribune*, 8
July 1889; *San Francisco Examiner*, 13 March 1892.

"tuneful tones of human progress" Oakland Merchants
Exchange and Oakland Board of Trade (1897), 67.

Opposition grew to outrage See Gary Kamiya's version
of the Gray Quarry story at https://www.sfgate.com/bayarea
/article/Quarrymen-hit-few-obstacles-to-blowing-up-hills
-5257200.php.

"The noise of traffic" Blake (1911), 256.

The Oakland Paving Company's former quarry See the
rocks at https://oaklandgeology.com/2020/07/06/rocks-of-the
-bilger-quarry/.

6 THE FAN, OR THE SECOND LEVEL

Sources on the Anza expeditions are made available by the University of Oregon at https://webdeanza.org/archives.html. My
interpretations of the Anza and Fages diaries do not agree in all
particulars with previous publications. The Fan is covered in detail in my *Oakland Geology* blog (https://oaklandgeology.com
/category/the-fan/). No one to my knowledge has attempted the
geological synthesis of facts I present in this chapter.

its outline resembles a tattered Japanese folding paper fan See the map at https://oaklandgeology.com /oakland-geologic-map/.

humans who arrived later turned the hills into open fields This was the pattern as Neanderthals moved into Europe; see Roebroecks et al. (2021).

a small reconnaissance mission See more at https:// oaklandgeology.com/2021/01/04/pedro-fages-and-the-oakland -fan/.

its southernmost hill See lobe 8 of the Fan at https:// oaklandgeology.com/2015/06/15/lobe-8-of-the-fan-evergreen -cemetery-hill/.

"very good and level country" Bolton (1911), 13.

"we were compelled to travel about a league and a half" Bolton (1927), 289.

Next to visit the Fan was Juan Bautista de Anza See more at https://oaklandgeology.com/2021/01/18/anza-and-the -fan/.

"very green and flower-strewn" Bolton (1930), 360.

the first building in their Rancho San Antonio See the site, now Peralta Hacienda Historical Park, at http://www .peraltahacienda.org/.

"compensated in a measure by the picturesque scenery" Scott (1871), 42.

Early photographs of the Brooklyn hills See one at https://oaklandgeology.com/2017/09/04/an-early-look-at-the -fan/.

"extinct late Pleistocene vertebrate fossils" Graymer (2000) is the latest to use the phrase.

it was accepted in the 1950s See Radbruch (1957).

the San Leandro gabbro See some at https://oakland geology.com/2017/02/20/two-bits-of-gabbro/.

7 INDIAN GULCH

For the history of Oakland during the Spanish and Mexican periods, see Bagwell (1982). The basics of streams and their characteristics are covered in chapter 11 of Johnson et al. (2017). The saga of Indian Gulch's park proposals was compiled from contemporary newspapers.

a map surveyed in 1853 See this map, made by Julius Kellersberger, at https://exhibits.stanford.edu/ruderman/catalog /xz539hd1973.

An 1869 map shows the glen's northern side See the map at https://www.lib.berkeley.edu/EART/maps/g4364-02 -1869-s4.html.

It got this name on 23 April 1893 *Oakland Tribune*, 24 April 1893.

"The most central and available location for a public park" *San Francisco Examiner*, 22 October 1896.

"It has sheltered nooks, a wooded glen" *Oakland Tribune*, 24 December 1897.

"We want a park, not a big ranch" *San Francisco Examiner*, 15 December 1897.

"Now the whole enterprise will be abandoned" San Francisco *Call*, 22 March 1898.

"a rugged canyon, of revolting appearance" San Francisco *Call*, 22 September 1904.

"a very retired and secluded locality" *Oakland Tribune*, 31 August 1888.

"more chances to acquire portions of Indian Gulch in 1914, 1917 and 1919" *Oakland Tribune*, 29 September 1915, 27 March 1917 and 28 October 1919.

"a spot where several human tragedies have taken place" *San Francisco Examiner*, 6 September 1909.

"unredeemable mistake" Hegemann (1915), 126.

8 THE BAY SHORE AND FLATS

The official history of Oakland Harbor (Minor 2000) is an invaluable source of specifics. Figuers (1998) describes the shallow structure and early water history of the East Bay in detail, and Noble and Montgomery (1999) is a deeply sourced history of East Bay MUD and its predecessors. Nelson (1909) is the primary source for shellmounds of the Bay Area.

"The greater part of the shore of the port" Teggart (1913), 69.

Jack London memorialized them Find London's short story "A Raid on the Oyster Pirates" at https://american literature.com/author/jack-london.

The geologic setting was the same ice-age marine terrace See the Clinton terrace at https://oaklandgeology .com/2017/03/20/the-marine-terrace-of-clinton-lengthwise/.

Horace Carpentier's greatest swindle See Halley (1876), 449–52.

a replica pier recycles vintage riprap from the 1870s See it at https://oaklandgeology.com/2011/07/05/replica -training-wall/.

The most ambitious one This was the so-called Reber plan; see https://en.wikipedia.org/wiki/Reber_Plan.

Much of it, or the ground beneath it, may soften or liquefy See a hazard map at https://earthquake.usgs.gov /hazards/urban/sfbay/liquefaction/alameda/.

"O, what a beautiful plain for a city" *Sacramento Bee*, 2 January 1868.

"the sea has eaten away at least 200 feet of the land" San Francisco *Call*, 10 September 1892.

bones of mastodons and giant ground sloths *Oakland Tribune*, 17 August 1964.

the Bay has had different shapes in the recent geologic past See Sarna-Wojcicki (2021).

East Bay MUD has begun storing water in aquifers south of San Leandro Bay Find details on the project, taking place at the Bayside Groundwater Facility, at https://www .ebmud.com.

land subsidence due to overpumping See more about this significant problem at https://www.usgs.gov/centers/land -subsidence-in-california.

9 SIBLEY VOLCANIC REGIONAL PRESERVE

Sarna-Wojcicki (2021) is an authoritative summary of the Miocene to Pleistocene paleogeography of the Bay Area. The Orinda and Moraga Formations are exhaustively treated by Wagner et al. (2021), including a detailed geologic map, and their displaced western counterparts likewise by Wagner et al. (2011).

Sibley Volcanic Regional Preserve is in a band of high land See basic park information at https://www.ebparks.org /parks/sibley/.

the Orinda Formation See examples in Claremont Canyon at https://oaklandgeology.com/2019/03/18 /claremont-canyon-conglomerate/.

These rules were among several principles stated by Nicolaus Steno See a summary at https://www.thoughtco.com /stenos-laws-or-principles-1440787.

"a machine of a peculiar construction" Hutton (1788), 209.

The regional park district took up this parcel in 2010 See East Bay Regional Park District (2018).

"untrammeled by man" The iconic phrase is from the 1964 Wilderness Act.

the Moraga Formation See a tour at https://www.kqed .org/quest/23196/geological-outings-around-the-bay-the -moraga-formation.

a "paleo" contractor sifted the tailings and recovered thousands of fossils See some at https://oaklandgeology .com/2012/11/23/rocks-of-the-caldecott-tunnel/; also Solon and McCosker (2014) and Risden and Morrison (2012).

The Miocene period is officially defined See these boundary points at https://stratigraphy.org/gssps/.

Absolute ages from the lava in the Moraga Formation Wagner et al. (2021) present the most recent data.

exposures show the tilted Orinda and Moraga Formations in perfect contact See it at https://oaklandgeology .com/2011/05/17/caldecott-tunnel-and-the-orinda-formation/.

"the rock is a fine-grained basalt" California State Mining Bureau (1906), 315.

"I decided to make a geologic map of the Preserve" Edwards (1983), 83.

those fossil bubbles later filled with mineral matter These are called amygdules; see some at https://oaklandgeology .com/2009/06/09/amygdules/.

white, porous freshwater limestone See it at https:// oaklandgeology.com/2019/07/22/oakland-geology-ramble-7/.

rearranging the greater East Bay for some twelve million years Graymer et al. (2002) detail this complex story.

10 LEONA HEIGHTS AND
THE SOUTHERN OAKLAND HILLS

The Ohlone ocher deposit was described by Wallace (1947). The mining history of Leona Heights is frustratingly vague. Official state accounts (California State Mining Bureau 1894, 1906; Mace 1911; Laizure 1929; Davis 1950) and academic papers (Schaller 1903, Clark 1917) are sketchy and inconsistent. As many as four mines have been active at Leona Heights, but the Alma and Leona Mines were predominant. Jennings and Strand (1963) list early graduate theses on California geology.

"a curious looking red baked clay or adobe" San Francisco *Call*, 10 September 1892.

Indigenous Californians have not forgotten this ancient trade See Oakland ocher at https://oaklandgeology .com/2013/05/24/oakland-ochre/.

"Old Survivor" See more at https://localwiki.org /oakland/Old_Survivor_Redwood_Tree.

"the Contra Costa Laundry" San Francisco *Call*, 2 December 1902.

"a vein of auriferous quartz" Cronise (1868), 154. There were isolated reports of gold-bearing quartz in the hills of Berkeley as well.

"trolley parties" *Oakland Tribune*, 19 June 1896.

a dip into plate tectonics Johnson et al. (2017) is a good online textbook of basic geology.

several different stories that account for the Leona Volcanics Dickinson, Hopson, and Saleeby (1996) compare three.

The following version was published in 2008 See Hopson et al. (2008).

Fritz Boehmer See Boehmer's colorful autobiography at https://calisphere.org/item/ark:/13030/kt8k4023t0/.

"a good article of domestic paint" Oakland Merchants Exchange and Oakland Board of Trade (1897), 23.

"I do not personally care to develop any mining property" *Alameda Daily Argus*, 22 July 1898.

he struck pyrite pay dirt *Oakland Tribune*, 23 June 1900.

produced about 160 tons of copper California Division of Mines (1948), 212.

"These natural vitriols" Schaller (1903), 192.

the walls of the former Crusher Quarry See them at https://oaklandgeology.com/2015/04/11/rocks-of-the-crusher -quarry/.

acid alteration has turned pockets of the rock into kaolinite See it at https://oaklandgeology.com/2016/01/25 /clay-outcrop-in-horseshoe-canyon/.

the great scar of the Leona Quarry See more at https:// oaklandgeology.com/2020/11/09/the-changing-identities -of-the-leona-quarry/.

streams and gutters are stained See etched and stained gutters at https://oaklandgeology.com/2014/07/27/geranium -place-rocks-and-runoff/.

the former Leona Heights Mine See the creekbed stains at https://oaklandgeology.com/2017/07/31/the-mine-drainage -of-leona-creek-revisited/.

"If the Leona Orebody is derived under vadose conditions" Mulryan (1925), 30.

I inspected a small horizontal tunnel See it at https:// oaklandgeology.com/2016/04/04/the-sulfur-problem/.

11 THE RIDGELINE

Marshall (2017) is a detailed sourcebook for the early history of the redwood belt. My main source for landslide science is Sidle and Ochiai (2006); and Nilsen, Taylor, and Brabb (1976) focus on Oakland landslides. The Great Valley Sequence, like the Franciscan Complex, is not easily summarized. Graymer (2000) is my go-to map of the bedrock in the Oakland Hills.

three tunnels, two of them long ago closed The two were the Kennedy Tunnel for cars (see https://localwiki.org /oakland/Kennedy_Tunnel) and the rail tunnel in Shepherd Canyon (see http://www.montclairrrtrail.org/railroad-to-trail -history.html).

"too conspicuous to be ignored" Beechey (1833).

landslides in Berkeley so large and gentle that people live on them Cohen-Waeber et al. (2018) look at these long-studied landslides.

the critical zone in the Oakland Hills has become drier Sidle and Ochiai (2006), 90–94.

the subduction factory Tatsumi (2005) is a basic introduction for geologists.

The California subduction zone varied over its lifetime Bentley et al. (2021) offer a case study based on Bay Area localities, "California's Coast Ranges: A Mesozoic Accretionary Wedge Complex."

bedrock maps of Oakland See Lawson (1914), Case (1963, 1968), Radbruch and Case (1967), Radbruch (1969) and Graymer (2000).

the Knoxville Formation See examples at https://oaklandgeology.com/2016/05/23/upper-arroyo-viejo-my-first-oakland-fossil/.

the Oakland Conglomerate See it at https://oakland geology.com/2018/06/11/oakland-geology-ramble-6/.

a mile of blue-green serpentine rock Geologists say *serpentinite*, accent on the *pent*, for the rock and *serpentine*, accent on the *serp*, for the mineral making it up. Wakabayashi (2017) is a deep survey of California's various serpentinites. See Oakland examples at https://oaklandgeology.com/2015/05/04/oaklands-serpentinite-patch/.

the Redwood Canyon Formation See outcrops at https://oaklandgeology.com/2018/02/19/redwood-ridge-and-the-parkridge-land-bridge/.

nameless mudstone of Eocene age See it at https://oaklandgeology.com/2019/04/01/the-eocene-mudstone-part-1/.

the subduction zone was in full swing My account is based mainly on Sharman et al. (2015) and Gooley, Grove, and Graham (2021).

a fat volcanic plateau rode into the trench See Liu et al. (2010). This plateau, the eastern counterpart to the Shatsky Rise in the western Pacific, now lies deep below the eastern seaboard.

the relative motion between the plates abruptly shifted See Tanya Atwater's pioneering animations at http://animations.geol.ucsb.edu/ and the treatment in Bentley et al. (2021).

the Sobrante Formation See this mudstone at https://oaklandgeology.com/2012/06/08/searching-for-the-sobrante/.

the steeply tilted Claremont Shale See a close-up at https://oaklandgeology.com/2015/12/21/shale-and-chert-in-the-claremont-shale/.

the geologist who mapped it could assign these rocks a Paleocene age Russ Graymer, written communication, 2009.

BIBLIOGRAPHY

Aldrich, Michele L., Bruce Bolt, Alan Leviton, and Peter Rodda. "The 'Report' of the 1868 Haywards Earthquake." *Bulletin of the Seismological Society of America 76,* no. 1 (1986): 71–76. https://doi.org/10.1785 /BSSA0760010071.

Bagwell, Elizabeth L. *Oakland, the Story of a City* (2nd ed.). Oakland: Oakland Heritage Alliance, 1982 (2012).

Baker, Joseph E. *Past and Present of Alameda County, California* (2 vols.). Chicago: S. J. Clarke Publishing, 1914. https://archive.org/details /pastpresentofala001bake.

Beechey, F. William. *Narrative of a Voyage to the Pacific and Beering's Strait, to Co-operate with the Polar Expeditions.* London: Henry Colburn and Richard Bentley, 1831. https://archive.org/details /narrativevoyage2beec.

Beechey, F. William. *The Harbour of San Francisco, Nueva California.* London: Hydrographical Office of the Admiralty, 1833. https://purl.stanford.edu /cm483ht8669.

Bentley, Callan, Karen Layou, Russ Kohrs, Shelley Jaye, Matthew D. Affolter, and Brian Ricketts. *Historical Geology, a Free Online Textbook for Historical Geology Courses.* Opengeology.org, 2021. https://opengeology .org/historicalgeology/.

Blake, Evarts I. *Greater Oakland 1911, a Volume Dealing with the Big Metropolis on the Shores of San Francisco Bay.* Oakland: Pacific Publishing, 1911. https://archive.org /details/greateroakland1901blak.

Boatwright, John, and Howard Bundock. *Modified Mercalli Intensity Maps for the 1868 Hayward Earthquake Plotted in ShakeMap Format.* Open-File Report 2008-1121. US Geological Survey, 2008. https://pubs.usgs.gov /of/2008/1121/.

Bolton, Herbert E., ed. "Expedition to San Francisco Bay in 1770: Diary of Pedro Fages." *[UC Berkeley] Academy of West*

Coast History Publications 2, no. 3 (1911). https://catalog.hathitrust.org/Record/101510420.

Bolton, Herbert E. *Fray Juan Crespi, Missionary Explorer on the Pacific Coast 1769–1774.* Berkeley: University of California Press, 1927. https://catalog.hathitrust.org/Record/000288788.

Bolton, Herbert E. *Anza's California Expeditions.* Vol. 4, *Font's Complete Diary of the Second Anza Expedition.* Berkeley: University of California Press, 1930. https://archive.org/details/anzascaliforniae04bolt/.

Buwalda, John P. "Nature of the Late Movements on the Haywards Rift, Central California." *Bulletin of the Seismological Society of America* 19, no. 4 (1929): 187–99. https://doi.org/10.1785/BSSA0190040187.

California Division of Mines. *Copper in California.* Bulletin 144. Sacramento: California Division of Mines, 1948. https://archiveorg/details/copperincaliforn00calirich/.

California State Mining Bureau. *Twelfth Report of the State Mineralogist (Second Biennial): Two Years Ending September 15, 1894.* Sacramento: California State Mining Bureau, 1894. https://archive.org/details/reportofstatemin12cali/.

California State Mining Bureau. *The Structural and Industrial Minerals of California.* Bulletin 38. Sacramento: California State Mining Bureau, 1906. https://archive.org/details/38calicturalindu00auburich/.

Case, James E. "Geology of a Portion of the Berkeley and San Leandro Hills, California." Ph.D. diss., University of California, Berkeley, 1963.

Case, James E. *Upper Cretaceous and Lower Tertiary Rocks, Berkeley and San Leandro Hills, California.* Bulletin 1251-J. US Geological Survey, 1968. https://doi.org/10.3133/b1251J.

Clark, Clifton W. "The Geology and Ore Deposits of the Leona Rhyolite." *University of California, Department of Geological Sciences Bulletin* 10, no. 20 (1917): 361–82. https://www.google.com/books/edition/University_of_California_Publications_in/UxxYAAAAYAAJ.

Cohen-Waeber, J., R. Bürgmann, E. Chaussard, C. Giannico, and

A. Ferretti. "Spatiotemporal Patterns of Precipitation-Modulated Landslide Deformation from Independent Component Analysis of InSAR Time Series." *Geophysical Research Letters* 45, no. 4 (2018): 1878–87. https://doi.org/10.1002/2017GL075950.

Conmy, Peter T. *The Beginnings of Oakland, California.* Oakland: Oakland Public Library, 1961. https://hdl.handle.net/2027/mdp.39015027940710.

Cronise, Titus Fey. *The Natural Wealth of California.* San Francisco: H. H. Bancroft, 1868. https://openlibrary.org/books/OL6936768M/The_natural_wealth_of_California.

Davis, Fenelon F. "Mines and Mineral Resources of Alameda County, California." *California Journal of Mines and Geology* 46, no. 2 (1950): 279–346. https://archive.org/details/californiajourna46cali/page/279.

Dickinson, William R., Clifford A. Hopson, and Jason B. Saleeby. "Alternate Origins of the Coast Range Ophiolite (California): Introduction and Implications." *GSA Today* 6, no. 2 (1996): 1–10. https://www.geosociety.org/gsatoday/archive/6/2/.

East Bay Regional Park District. *Robert Sibley Volcanic Regional Preserve Land Use Plan Amendment, Incorporating the McCosker Parcel and Western Hills Open Space.* Oakland: East Bay Regional Park District, 2018.

Edwards, Stephen W. "Ancient Volcanic Features of the Berkeley Hills." *California Geology* 36, no. 4 (1983): 83–87. https://filerequest.conservation.ca.gov/RequestFile/57635.

Elliott, W. W. *Oakland and Surroundings Illustrated and Described Showing Its Advantages for Residence or Business.* Oakland: W. W. Elliott Publisher, 1885. https://openlibrary.org/works/OL16766147W/Oakland_and_surroundings.

Evanosky, Dennis. *Oakland's Mountain View Cemetery.* Alameda, CA: Stellar Media Group, 2007.

Figuers, Sands H. *Groundwater Study and Water Supply History of the East Bay Plain, Alameda and Contra Costa*

Counties, CA. Livermore, CA: Norfleet Consultants, prepared for Friends of the San Francisco Estuary, 1998. https://www.waterboards.ca.gov/sanfranciscobay /water_issues/programs/groundwater/groundwaterstudy .html.

Gooley, Jared T., Marty Grove, and Stephan A. Graham. "Tectonic Evolution of the Central California Margin as Reflected by Detrital Zircon Composition in the Mount Diablo Region." In *Regional Geology of Mount Diablo, California: Its Tectonic Evolution on the North America Plate Boundary*, edited by Raymond Sullivan, Doris Sloan, Jeffrey R. Unruh, and David P. Schwartz, 305–29. Memoir 217. Boulder, CO: Geological Society of America, 2021. https://doi.org/10.1130/2021.1217(14).

Graymer, Russ W. *Geologic Map and Map Database of the Oakland Metropolitan Area, Alameda, Contra Costa, and San Francisco Counties, California*. Miscellaneous Field Studies Map MF-2342. US Geological Survey, 2000. https://doi.org/10.3133/mf2342.

Graymer, R. W., A. M. Sarna-Wojcicki, J. P. Walker, R. J. McLaughlin, and R. J. Fleck. "Controls on Timing and Amount of Right-Lateral Offset on the East Bay Fault System, San Francisco Bay Region, California." *Geological Society of America Bulletin* 114, no. 12 (2002): 1471–79. https://doi.org/10.1130/0016-7606(2002)114<1471: COTAAO>2.0.CO;2.

Guinn, James M. *History of the State of California and Biographical Record of Oakland and Environs, Also Containing Biographies of Well-Known Citizens of the Past and Present*. Los Angeles: Historic Record, 1907. https://archive.org/details/historyofstateofooguin.

Halley, William. *The Centennial Year Book of Alameda County, California*. Oakland: William Halley, 1876. https:// archive.org/details/centennialyearbooohall.

Hansen, Jens Morten. "On the Origin of Natural History: Steno's Modern, but Forgotten Philosophy of Science." *Bulletin of the Geological Society of Denmark* 57 (2009): 1–24. https://2dgf.dk/xpdf/bull57-1-24.pdf.

Harris, Ruth A., M. Barall, D. A. Lockner, D. E. Moore, D. A. Ponce, R. W. Graymer, G. Funning, C. A. Morrow, C. Kyriaka-poulos, and D. Eberhart-Phillips. "A Geology and Geod-esy Based Model of Dynamic Earthquake Rupture on the Rodgers Creek–Hayward–Calaveras Fault System, California." *Journal of Geophysical Research Solid Earth* 126, no. 3 (2021). https://doi.org/10.1029 /2020JB020577.

Hegemann, Werner. *Report on a City Plan for the Municipalities of Oakland & Berkeley*. Municipal Governments of Oakland and Berkeley et al., 1915. https://archive.org /details/reportoncityplan00hegerich.

Helley, Edward J., John A. Fitzpatrick, and James L. Bischoff. *Uranium-Series Dates on Oyster Shells from Marine Terraces of San Pablo Bay, California*. Open-File Report 93-286. US Geological Survey, 1993. https://doi .org/10.3133/ofr93286.

Hopson, Clifford A., James M. Mattinson, Emile A. Pessagno Jr., and Bruce P. Luyendyk. "California Coast Range Ophiolite: Composite Middle and Late Jurassic Oceanic Lithosphere." In *Ophiolites, Arcs, and Batholiths: A Tribute to Cliff Hopson*, edited by J. E. Wright and J. W. Shervais, 1–101. Special Paper 438. Boulder, CO: Geological Society of America, 2008. https://doi .org/10.1130/2008.2438(01).

Hough, Susan E., and Roger G. Bilham. *After the Earth Quakes: Elastic Rebound on an Urban Planet*. New York: Oxford University Press, 2006.

Hutton, James. "Theory of the Earth, or an Investigation of the Laws Observable in the Composition, Dissolution, and Restoration of Land upon the Globe." *Transactions of the Royal Society of Edinburgh* 1, no. 2 (1788): 209–304. https://doi.org/10.1017/S0080456800029227.

Jennings, Charles W., and Rudolph G. Strand. *Index to Graduate Theses on California Geology to December 31, 1961*. Special Report 74. San Francisco: California Division of Mines and Geology, 1963. https://archive.org/details /indextograduatet74jenn.

Johnson, Chris, Matthew D. Affolter, Paul Inkenbrandt, and Cam
    Mosher. *An Introduction to Geology: A Free Textbook
    for College-Level Introductory Geology Courses.*
    Opengeology.org, 2017. https://opengeology.org
    /textbook/.

Jones, Richard Foster, ed. *Francis Bacon: Essays, Advancement of
    Learning, New Atlantis, and Other Pieces.* New York:
    Odyssey Press, 1937.

Jordan, David Starr. "The Earthquake Rift of April, 1906." In
    The California Earthquake of 1906, edited by David
    Starr Jordan, 1–62. San Francisco: A. M. Robertson,
    1907. https://oac.cdlib.org/view?docId=hb7r29p21q.

Laizure, C. McK. "Alameda County." In *Report XXV of the State
    Mineralogist Covering Mining in California and
    Activities of the Division of Mines,* edited by Walter W.
    Bradley, 427–56. Sacramento: Division of Mines, 1929.
    https://archive.org/details/chapterofreport025cali
    /page/427/.

Lawson, Andrew C. *The California Earthquake of April 18, 1906:
    Report of the State Earthquake Investigation Commis-
    sion.* Publication 87. Washington: Carnegie Institute of
    Washington, 1908. https://calisphere.org/item
    /ark:/13030/hb1h4n989f/.

Lawson, Andrew C. "Description of the San Francisco District."
    In *San Francisco Folio, California, Tamalpais, San
    Francisco, Concord, San Mateo, and Haywards Quad-
    rangles.* Geologic Atlas Folio 193. US Geological Survey,
    1914. https://doi.org/10.3133/gf193.

Le Conte, Joseph. *Elements of Geology.* New York: D. Appleton and
    Company, 1879. https://archive.org/details
    /elementsgeo00lecorich.

Lienkaemper, James J. *Digital Database of Recently Active Traces
    of the Hayward Fault, California.* Data Series 177. US
    Geological Survey, 2006. https://doi.org/10.3133/ds177.

Lienkaemper, James J., and Patrick L. Williams. "Evidence for
    Surface Rupture in 1868 on the Hayward Fault in North
    Oakland and Major Rupturing in Prehistoric Earth-
    quakes." *Geophysical Research Letters* 26, no. 13 (1999):

1949–52. http://dx.doi.org/10.1029/1999GL900393.

Lienkaemper, James J., Patrick L. Williams, and T. P. Guilderson. "Evidence for a Twelfth Large Earthquake on the Southern Hayward Fault in the Past 1900 Years." *Bulletin of the Seismological Society of America* 100, no. 5a (2010): 2024–34. https://doi.org/10.1785/0120090129.

Liu, Lijun, Michael Gurnis, Maria Seton, Jason Saleeby, R. Dietmar Müller, and Jennifer M. Jackson. "The Role of Oceanic Plateau Subduction in the Laramide Orogeny." *Nature Geoscience* 3 (2010): 353–57. https://doi.org/10.1038/ngeo829.

Louderback, George D. "Central California Earthquakes of the 1830's." *Bulletin of the Seismological Society of America* 37, no. 1 (1947): 33–74. https://doi.org/10.1785/BSSA0370010033.

Mace, Clement H. "Genesis of Leona Heights Ore Deposit, California." *Mining and Engineering World* 35 (1911): 1320.

Marshall, Amelia Sue. *East Bay Hills, a Brief History*. Charleston, SC: History Press, 2017.

McFarland, F. S., J. J. Lienkaemper, and S. J. Caskey. *Data from Theodolite Measurements of Creep Rates on San Francisco Bay Region Faults, California, Version 1.8, March 2016*. Open-File Report OF2009-1119. US Geological Survey, 2016. https://doi.org/10.3133/ofr20091119.

Merritt, Frank C. *History of Alameda County, California* (2 vols.). Chicago: S. J. Clark Publishing, 1928. https://archive.org/details/historyofalameda01merr/, https://archive.org/details/historyofalameda02merr/.

Minor, Woodruff C. *Pacific Gateway, an Illustrated History of the Port of Oakland*. Oakland: Port of Oakland, 2000.

Mulryan, Henry. "Geology and Ores of the Leona Heights Pyrite Deposit, Oakland, California." Master's diss., Stanford University, 1925. https://searchworks.stanford.edu/view/1171838.

Nelson, N. C. "Shellmounds of the San Francisco Bay Region." *University of California Publications in American Archaeology and Ethnology* 7, no. 4 (1909): 309–48.

https://digitalassets.lib.berkeley.edu/anthpubs/ucb/text/ucp007-006-007.pdf.

Nilsen, T. H., F. A. Taylor, and E. E. Brabb. *Recent Landslides in Alameda County, California (1940–71): An Estimate of Economic Losses and Correlations with Slope, Rainfall, and Ancient Landslide Deposits*. Bulletin 1398. US Geological Survey, 1976. https://doi.org/10.3133/b1398.

Noble, John Henry, and Gayle B. Montgomery. *Its Name Was M.U.D., a Story of Water*. Oakland: East Bay Municipal Utility District, 1999.

Oakland Merchants Exchange and Oakland Board of Trade. *Oakland, Athens of the Pacific, Facts and Figures of Oakland and Alameda County*. Oakland: G. T. Loofbourow & Co., 1897. https://archive.org/details/oaklandathensofpoocald.

Phelps, G. A., R. W. Graymer, R. C. Jachens, D. A. Ponce, R. W. Simpson, and C. M. Wentworth. *Three-Dimensional Geologic Map of the Hayward Fault Zone, San Francisco Bay Region, California*. Scientific Investigations Map 3045. US Geological Survey, 2008. https://doi.org/10.3133/sim3045.

Prescott, William H. "Circumstances Surrounding the Preparation and Suppression of a Report on the 1868 California Earthquake." *Bulletin of the Seismological Society of America* 72, no. 6a (1982): 2389–93. https://doi.org/10.1785/BSSA07206A2389.

Radbruch, Dorothy H. *Areal and Engineering Geology of the Oakland West Quadrangle, California*. Miscellaneous Investigations Map I-239. US Geological Survey, 1957. https://doi.org/10.3133/i239.

Radbruch, Dorothy H. *Areal and Engineering Geology of the Oakland East Quadrangle, California*. Map GQ-769. US Geological Survey, 1969. https://doi.org/10.3133/gq769.

Radbruch, Dorothy H., and J. E. Case. *Preliminary Geologic and Engineering Geologic Information, Oakland and Vicinity, California*. Open-File Report 67-183. US Geological Survey, 1967. https://doi.org/10.3133/ofr67183.

Raymond, Loren A. "What Is Franciscan? Revisited." *International*

Geology Review 60 (2017): 1968–2030. https://doi.org
/10.1080/00206814.2017.1396933.

Risden, Chris, and Ivy Morrison. *The Caldecott Fourth Bore Project: Tunneling through a Miocene Plate Boundary*. Northern California Geological Society field trip guide, 2012. http://www.ncgeolsoc.org/wp-content/uploads /2020/02/2012-3_Caldecott-Tunnel-Fourth-Bore_MAS-TER.pdf.

Roebroecks, Wil, Katharine MacDonald, Fulco Scherjon, Corrie Bakels, Anastasia Nikulina, Eduard Pop, Lutz Kindler, and Sabine Gaudzinski-Windheuser. "Landscape Modification by Last Interglacial Neanderthals." *Science Advances* 7 (2021), no. 51. https://doi.org /10.1126/sciadv.abj5567.

Rowlandson, Thomas. *A Treatise on Earthquake Dangers, Causes and Palliatives*. San Francisco: Dewey & Co. Publishers, 1869. https://archive.org/details/treatiseonearthqoorowl.

Sarna-Wojcicki, Andrei M. "Late Cenozoic Paleogeographic Reconstruction of the San Francisco Bay Area from Analysis of Stratigraphy, Tectonics, and Tephrochronology." In *Regional Geology of Mount Diablo, California: Its Tectonic Evolution on the North America Plate Boundary*, edited by Raymond Sullivan, Doris Sloan, Jeffrey R. Unruh, and David P. Schwartz, 443–72. Memoir 217. Boulder, CO: Geological Society of America, 2021. https://doi.org/10.1130/2021.1217(17).

Schaller, Waldemar T. "Minerals from Leona Heights, Alameda Co., California." *[University of California] Bulletin of the Department of Geology* 3, no. 7 (1903): 191–217. https://www.google.com/books/edition/Euceratherium/ IEUxAQAAMAAJ?gbpv=1.

Schwarzer, Mitchell. *Hella Town: Oakland's History of Development and Disruption*. Oakland: University of California Press, 2021.

Scott, John. *Information Concerning the Terminus of the Railroad System of the Pacific Coast*. Oakland: Daily Transcript, 1871. https://archive.org/details/informationconco1 scotgoog.

Sharman, Glenn R., Stephan A. Graham, Marty Grove, David L. Kimbrough, and James E. Wright. "Detrital Zircon Provenance of the Late Cretaceous–Eocene California Forearc: Influence of Laramide Low-Angle Subduction on Sediment Dispersal and Paleogeography." *Geological Society of America Bulletin* 127, no. 1/2 (2015): 38–60. https://doi.org/10.1130/B31065.1.

Sidle, R. C., and H. Ochiai. *Landslides: Processes, Prediction, and Land Use*. Water Resources Monograph 18. Washington, DC: American Geophysical Union, 2006. https://doi.org/10.1029/WM018.

Solon, Mary, and Mary McCosker. *Building the Caldecott Tunnel*. Charleston, SC: Arcadia Publishing, 2014.

Tatsumi, Yoshiyuki. "The Subduction Factory: How It Operates in the Evolving Earth." *GSA Today* 17, no. 7 (2005): 4–10. https://www.geosociety.org/gsatoday/archive/15/7/pdf /i1052-5173-15-7-4.pdf.

Teggart, Frederick J., ed. *The Anza Expedition of 1775–1776: Diary of Pedro Font*. Berkeley: University of California Publications of the Academy of Pacific Coast History, 1913. https://archive.org/details/anzaexpeditionof00font.

Toké, Nathan A., Christopher G. Boone, and J. Ramón Arrowsmith. "Fault Zone Regulation, Seismic Hazard, and Social Vulnerability in Los Angeles, California: Hazard or Urban Amenity?" *Earth's Future* 2, no. 9 (2014): 440–57. https://doi.org/10.1002/2014EF000241.

Wagner, David L., George J. Saucedo, Kevin B. Clahan, Robert J. Fleck, Victoria E. Langenheim, Robert J. McLaughlin, Andrei M. Sarna-Wojcicki, James R. Allen, and Alan L. Deino. "Geology, Geochronology, and Paleogeography of the Southern Sonoma Volcanic Field and Adjacent Areas, Northern San Francisco Bay Region, California." *Geosphere* 7, no. 3 (2011): 658–83. https://doi.org /10.1130/GES00626.1.

Wagner, J. Ross, Alan Deino, Stephen W. Edwards, Andrei M. Sarna-Wojcicki, and Elmira Wan. "Miocene Stratigraphy and Structure of the East Bay Hills, California." In *Regional Geology of Mount Diablo, California: Its Tectonic Evolution on the North America Plate Boundary*, edited by Raymond Sullivan, Doris Sloan, Jeffrey R. Unruh, and David P. Schwartz, 331–91. Memoir 217. Boulder, CO: Geological Society of America, 2021. https://doi .org/10.1130/2021.1217(15).

Wakabayashi, John. "Anatomy of a Subduction Complex: Architecture of the Franciscan Complex, California, at Multiple Length and Time Scales." *International Geology Review* 57 (2015): 669–746. https://doi.org/10.1080 /00206814.2014.998728.

Wakabayashi, John. "Serpentinites and Serpentinites: Variety of Origins and Emplacement Mechanisms of Serpentinite Bodies in the California Cordillera." *Island Arc* 26, no. 5 (2017). https://doi.org/10.1111/iar.12205.

Wallace, William J. "An Aboriginal Hematite Quarry in Oakland, California." *American Antiquity* 12, no. 4 (1947): 272–73. https://doi.org/10.2307/275058.

Walters, Mason. "The Seismic Retrofit of the Oakland City Hall." In *SMIP03 Seminar on Utilization of Strong-Motion Data Proceedings*, edited by Moh Huang, 149–63. Sacramento: California Geological Survey, 2003. https:// www.conservation.ca.gov/cgs/Documents /Program-SMIP/Seminar/SMIP03/Paper10_Walters.pdf.

Waring, Gerald A. *Springs of California*. Water-Supply Paper 338. US Geological Survey, 1915. https://doi.org/10.3133 /wsp338.

Wendte, Charles William. *The Wider Fellowship: Memories, Friendships, and Endeavors for Religious Unity, 1844– 1927*. Vol. 2. Boston: Beacon Press, 1927. https:// babel.hathitrust.org/cgi/pt?id=mdp.39015048711132.

Wood, Myron W., and J. P. Munro-Fraser. *History of Alameda County, California*. Oakland: Pacific Press, 1883. https://archive.org/details/cu31924028881188/.

INDEX

ABOUT THE AUTHOR

Andrew Alden is a geologist and geoscience writer who has worked for the US Geological Survey and reported for KQED and *Bay Nature*. Long fascinated with rocks and landscapes, Alden found inspiration for his debut book, *Deep Oakland*, in the 1989 Loma Prieta earthquake, which, as he writes, "ripped the city open and revealed to us its heart and character." Through his writing Alden raises awareness for what he calls the deep present: the appreciation of the ancient underpinnings that shape the modern-day surroundings of daily life. His website is oaklandgeology.com.